Nadia Nedjah, Leandro Santos Coelho, Viviana Cocco Mariani,
and Luiza de Macedo Mourelle (Eds.)

Innovative Computing Methods and Their Applications to Engineering Problems

Studies in Computational Intelligence, Volume 357

Editor-in-Chief
Prof. Janusz Kacprzyk
Systems Research Institute
Polish Academy of Sciences
ul. Newelska 6
01-447 Warsaw
Poland
E-mail: kacprzyk@ibspan.waw.pl

Further volumes of this series can be found on our homepage: springer.com

Vol. 334. Álvaro Herrero and Emilio Corchado (Eds.)
Mobile Hybrid Intrusion Detection, 2011
ISBN 978-3-642-18298-3

Vol. 335. Radomir S. Stankovic and Radomir S. Stankovic
From Boolean Logic to Switching Circuits and Automata, 2011
ISBN 978-3-642-11681-0

Vol. 336. Paolo Remagnino, Dorothy N. Monekosso, and Lakhmi C. Jain (Eds.)
Innovations in Defence Support Systems – 3, 2011
ISBN 978-3-642-18277-8

Vol. 337. Sheryl Brahnam and Lakhmi C. Jain (Eds.)
Advanced Computational Intelligence Paradigms in Healthcare 6, 2011
ISBN 978-3-642-17823-8

Vol. 338. Lakhmi C. Jain, Eugene V. Aidman, and Canicious Abeynayake (Eds.)
Innovations in Defence Support Systems – 2, 2011
ISBN 978-3-642-17763-7

Vol. 339. Halina Kwasnicka, Lakhmi C. Jain (Eds.)
Innovations in Intelligent Image Analysis, 2010
ISBN 978-3-642-17933-4

Vol. 340. Heinrich Hussmann, Gerrit Meixner, and Detlef Zuehlke (Eds.)
Model-Driven Development of Advanced User Interfaces, 2011
ISBN 978-3-642-14561-2

Vol. 341. Stéphane Doncieux, Nicolas Bredeche, and Jean-Baptiste Mouret(Eds.)
New Horizons in Evolutionary Robotics, 2011
ISBN 978-3-642-18271-6

Vol. 342. Federico Montesino Pouzols, Diego R. Lopez, and Angel Barriga Barros
Mining and Control of Network Traffic by Computational Intelligence, 2011
ISBN 978-3-642-18083-5

Vol. 343. Kurosh Madani, António Dourado Correia, Agostinho Rosa, and Joaquim Filipe (Eds.)
Computational Intelligence, 2011
ISBN 978-3-642-20205-6

Vol. 344. Atilla Elçi, Mamadou Tadiou Koné, and Mehmet A. Orgun (Eds.)
Semantic Agent Systems, 2011
ISBN 978-3-642-18307-2

Vol. 345. Shi Yu, Léon-Charles Tranchevent, Bart De Moor, and Yves Moreau
Kernel-based Data Fusion for Machine Learning, 2011
ISBN 978-3-642-19405-4

Vol. 346. Weisi Lin, Dacheng Tao, Janusz Kacprzyk, Zhu Li, Ebroul Izquierdo, and Haohong Wang (Eds.)
Multimedia Analysis, Processing and Communications, 2011
ISBN 978-3-642-19550-1

Vol. 347. Sven Helmer, Alexandra Poulovassilis, and Fatos Xhafa
Reasoning in Event-Based Distributed Systems, 2011
ISBN 978-3-642-19723-9

Vol. 348. Beniamino Murgante, Giuseppe Borruso, and Alessandra Lapucci (Eds.)
Geocomputation, Sustainability and Environmental Planning, 2011
ISBN 978-3-642-19732-1

Vol. 349. Vitor R. Carvalho
Modeling Intention in Email, 2011
ISBN 978-3-642-19955-4

Vol. 350. Thanasis Daradoumis, Santi Caballé, Angel A. Juan, and Fatos Xhafa (Eds.)
Technology-Enhanced Systems and Tools for Collaborative Learning Scaffolding, 2011
ISBN 978-3-642-19813-7

Vol. 351. Ngoc Thanh Nguyen, Bogdan Trawiński, and Jason J. Jung (Eds.)
New Challenges for Intelligent Information and Database Systems, 2011
ISBN 978-3-642-19952-3

Vol. 352. Nik Bessis and Fatos Xhafa (Eds.)
Next Generation Data Technologies for Collective Computational Intelligence, 2011
ISBN 978-3-642-20343-5

Vol. 353. Igor Aizenberg
Complex-Valued Neural Networks with Multi-Valued Neurons, 2011
ISBN 978-3-642-20352-7

Vol. 354. Ljupco Kocarev and Shiguo Lian (Eds.)
Chaos-Based Cryptography, 2011
ISBN 978-3-642-20541-5

Vol. 355. Yan Meng and Yaochu Jin (Eds.)
Bio-Inspired Self-Organizing Robotic Systems, 2011
ISBN 978-3-642-20759-4

Vol. 356. Slawomir Koziel and Xin-She Yang (Eds.)
Computational Optimization, Methods and Algorithms, 2011
ISBN 978-3-642-20858-4

Vol. 357. Nadia Nedjah, Leandro Santos Coelho, Viviana Cocco Mariani, and Luiza de Macedo Mourelle (Eds.)
Innovative Computing Methods and Their Applications to Engineering Problems, 2011
ISBN 978-3-642-20957-4

Nadia Nedjah, Leandro Santos Coelho,
Viviana Cocco Mariani, and
Luiza de Macedo Mourelle (Eds.)

Innovative Computing Methods and Their Applications to Engineering Problems

Springer

Prof. Nadia Nedjah
State University of Rio de Janeiro
Department of Electronics Engineering and Telecommunications
Faculty of Engineering
Brazil
E-mail: nadia@eng.uerj.br
Tel.: +55 21 2334 0105 (Ext. 39)
Fax: +55 21 2334 0306

Dr. Leandro dos Santos Coelho
Pontifícia Universidade
Católica do Paraná
Centro de Ciências
Rua Imaculada Conceição 1155
80215-901 Curitiba
Brazil
E-mail: leandro.coelho@pucpr.br

Dr. Viviana Cocco Mariani
Pontifical Catholic University of Parana
Mechanical Engineering Graduate Program
Imaculada Conceição, 1155
80215-901 Curitiba
Brazil
E-mail: viviana.mariani@pucpr.br

Dr. Luiza de Macedo Mourelle
Universidade do Estado do Rio de Janeiro
Faculdade de Engenharia
Rua São Francisco Xavier 524
Sala 5022
20559-900 Rio de Janeiro
Brazil
E-mail: ldmm@ eng.uerj.br

ISBN 978-3-642-26836-6 ISBN 978-3-642-20958-1 (eBook)

DOI 10.1007/978-3-642-20958-1

Studies in Computational Intelligence ISSN 1860-949X

Softcover reprint of the hardcover 1st edition 2011

Typeset & Cover Design: Scientific Publishing Services Pvt. Ltd., Chennai, India.

Printed on acid-free paper

9 8 7 6 5 4 3 2 1

springer.com

Preface

The design of most modern engineering systems entails the consideration of a good trade-off between the several targets requirements to be satisfied along the system life such as high reliability, low redundancy and low operational costs. These aspects are often in conflict with one another, hence a compromise solution has to be sought. Innovative computing techniques, such as genetic algorithms, swarm intelligence, differential evolution, multi-objective evolutionary optimization, just to name few, are of great help in founding effective and reliable solution for many engineering problems.

Each chapter of this book attempts to using an innovative computing technique in solving a different engineering problem. The contributions of each and every one are summarized in the following paragraph.

In Chapter 1, the authors approach the traveling salesman using the discrete differential evolution approach. In Chapter 2, the authors solve reliability optimization problems of series-parallel, parallel-series and complicated system, using genetic algorithms. In Chapter 3, the authors present exploit particle swarm optimization to build fuzzy systems automatically. In Chapter 4, the authors propose a maintenance optimization of wind turbine systems using intelligent prediction tools. In Chapter 5, the authors apply the clonal selection algorithm to economic dispatch optimization of electrical energy. In Chapter 6, the authors present novel methods for multi-objective evolutionary optimization based on a dynamic aggregation of objectives. In Chapter 7, the authors solve the problems related to application mapping into a network-on-chip platform, using multi-objective evolutionary optimization. In Chapter 8, the authors introduce the theory of chaotic optimization methods together with some applications.

The editors are very much grateful to the authors of this volume and to the reviewers for their tremendous service by critically reviewing the chapters. The editors would like also to thank Prof. Janusz Kacprzyk, the editor-in-chief of the Studies in Computational Intelligence Book Series and Dr. Thomas Ditzinger, Springer Verlag, Germany for the editorial assistance and excellent cooperative collaboration to produce this important scientific work. We hope that the reader will share our excitement to present this volume and will find it useful.

March 2011

Nadia Nedjah
State University of Rio de Janeiro, Brazil

Leandro S. Coelho
Pontifical Catholic University of Parana, Brazil

Viviana C. Mariani
Pontifical Catholic University of Parana, Brazil

Luiza M. Mourelle
State University of Rio de Janeiro, Brazil

Contents

A Discrete Differential Evolution Approach with Local Search for Traveling Salesman Problems

João Guilherme Sauer[1], Leandro dos Santos Coelho[1,3], Viviana Cocco Mariani[2,3], Luiza de Macedo Mourelle[4], and Nadia Nedjah[5]

[1] Industrial and Systems Engineering Graduate Program, PPGEPS
[2] Mechanical Engineering Graduate Program, PPGEM
Pontifical Catholic University of Parana, PUCPR
Imaculada Conceição, 1155, Zip code 80215-901, Curitiba, PR, Brazil
{viviana.mariani,leandro.coelho}@pucpr.br,
joao.sauer@gmail.com
[3] Department of Electrical Engineering, PPGEE
Federal University of Parana, UFPR
Polytechnic Center, Zip code 81531-970, Curitiba, Parana, Brazil
[4] Department of Systems Engineering and Computation
Faculty of Engineering, State University of Rio de Janeiro
Rua São Francisco Xavier, 524, Zip code 20550-900, Rio de Janeiro, RJ, Brazil
ldmm@eng.uerj.br
[5] Department of Electronics Engineering and Telecommunications
Faculty of Engineering, State University of Rio de Janeiro
Rua São Francisco Xavier, 524, Zip code 20550-900, Rio de Janeiro, RJ, Brazil
{ldmm,nadia}@eng.uerj.br

Abstract. Combinatorial optimization problems are very commonly seen in scientific research and practical applications. Traveling Salesman Problem (TSP) is one nonpolynomial-hard combinatorial optimization problem. It can be describe as follows: a salesman, who has to visit clients in different cities, wants to find the shortest path starting from his home city, visiting every city exactly once and ending back at the starting point. There are exact algorithms, such as cutting-plane or facet-finding, are very complex and demanding of computing power to solve TSPs. There here, however, metaheuristics based on evolutionary algorithms that are useful to finding solutions for a much wider range of optimization problems including the TSP. Differential Evolution (DE) is a relatively simple evolutionary algorithm, which is an effective adaptive approach to global optimization over continuous search spaces. Furthermore, DE has emerged as one of the fast, robust, and efficient global search heuristics of current interest. DE shares similarities with other evolutionary algorithms, it differs significantly in the sense that distance and direction information from the current population is used to guide the

N. Nedjah et al. (Eds.): Innovative Computing Methods, SCI 357, pp. 1–12.
springerlink.com

search process. Since its invention, DE has been applied with success on many numerical optimization problems outperforming other more popular metaheuristics such as the genetic algorithms. Recently, some researchers extended with success the application of DE to combinatorial optimization problems with discrete decision variables. In this paper, the following discrete DE approaches for the TSP are proposed and evaluated: i) DE approach without local search, ii) DE with local search based on Lin-Kernighan-Heulsgaun (LKH) method, and iii) DE with local search based on Variable Neighborhood Search (VNS) and together with LKH method. Numerical study is carried out using the TSPLIB of test TSP problems. In this context, the computational results are compared with the other results in the recent TSP literature. The obtained results show that LKH method is the best method to reach optimal results for TSPLIB benchmarks, but for largest problems, the DE+VNS improve the quality of obtained results.

Keywords: Optimization, Traveling salesman problem, Evolutionary Algorithm, Differential Evolution, Variable Neighbor Search, Local Search, Lin-Kernighan-Heulsgaun.

1 Introduction

Combinatorial optimization problems occur in various fields of physics, engineering and economics. Many of them are difficult to solve since they are nonpolynomial-hard (NP-hard), i.e., there is no known algorithm that finds the exact solution with an effort proportional to any power of the problem size.

The Traveling Salesman Problem (TSP) is a well-known example of a NP-Hard combinatorial optimization problem that involves finding the shortest Hamiltonian cycle in a complete graph of n nodes (cities). In the TSP, it is given n cities 1, 2,…, n together with all the pair wise distances d_{ij} between cities i and j. The goal is to find the shortest tour that visits every city exactly once and in the end returns to its starting city.

TSPs raise important issues because many problems and practical applications in science, engineering, and bioinformatics fields, such as vehicle routing problems, integrated circuit board chip insertion problems, scheduling problems, flexible manufacturing systems, printed circuit board, X-ray crystallography, punching sequence problems, routing, job scheduling problems, and phylogenetic tree construction can be formulated as TSPs.

TSP has been extensively studied. A large number of approaches have been developed for solving TSPs (see [1]-[8]). In this context, there is a great interest in efficient procedures based on heuristics and metaheuristics to solve it. The TSP has received considerable attention over the last two decades and various approaches are proposed to solve the problem, such as branch-and-bound, cutting planes, 2-*opt*, 3-*opt*, simulated annealing, artificial neural network, and taboo search [1],[2]. Some of these methods are exact algorithms, while the others are near-optimal or approximate algorithms [3]-[5]. Recently, evolutionary algorithm approaches are successfully implemented to the TSP [6]-[8]. In other hand, evolutionary algorithms perform a typically incomplete search in the space of solutions by iteratively creating and evaluating new candidate solutions.

Differential Evolution (DE) is a relatively new evolutionary algorithm, which is an effective adaptive approach to global optimization over continuous search spaces. DE as developed by Storn and Price [9] has proven to be a promising candidate to solve real-valued optimization problems [10]. DE has emerged as one of the fast, robust, and efficient global search heuristics of current interest. The computational algorithm of DE is very simple and easy to implement, with only a few parameters required to be set by a user.

Classical DE approaches use the floating-point (real-coded) representation randomly generated initial population, differential mutation, probability crossover and greedy criterion of search.

The applications of DE on combinatorial optimization problems are still considered limited, but the advantages of DE include a simple structure, speed to acquire solutions, and robustness that are sustained in the literature. However, the major obstacle of successfully applying a DE algorithm to combinatorial problems in the literature is due to its continuous nature [11]. Aiming at the discrete problems, novel discrete DE approaches have been proposed in recent literature to solve combinatorial optimization problems [12]-[15]. In other hand, in spite of the prominent merits, sometimes DES shows the premature convergence and slowing down of convergence as the region of global optimum is approached. The application of local search in DE is an alternative strategy to improve the convergence performance.

In this paper, the following discrete DE approaches for the TSP are proposed and evaluated: i) discrete DE approach without local search, ii) DE with local search based on Lin-Kernighan-Heulsgaun (LKH) [16] method, and iii) DE with local search based on Variable Neighborhood Search (VNS) [17],[18] and together with LKH method. Computational results evaluated in the TSP are based on Reinelt's TSPLIB [19]. In this context, those results are compared with the other results in the recent TSP literature. The obtained results show that LKH method is the best method to reach optimal results for TSPLIB benchmarks, but for largest problems, the DE+VNS improve the quality of obtained results.

In general terms, the contribution of this paper was the application of an extension of DE algorithm using discrete variables and local search mechanisms to a set of benchmark problems described in TSPLIB and studied the incorporation of different local search schemes to improve the performance of discrete DE. Simulation results have been presented to compare the performance of different schemes.

The remainder of this paper is organized as follows. In Section 2, a describtion of the fundamentals of TSP is provided. Section 3 presents the features of discrete DE approach with local search. Section 4 then describes the TSPLIB benchmark problems and evaluates the quality of simulation results. Lastly, section 5 presents our conclusion and future research.

2 Fundamentals of TSP

The TSP is one of the most extensively studied problems in combinatorial optimization. Problems of combinatorial optimization distinguish themselves by their well-structured problem description as well as by their huge number of possible action alternatives.

The task of TSP basically consists of finding the shortest tour through a number of cities, visiting every city exactly once. The TSP can be formulated as follows. Given matrix $D=(d_{ij})_{n\times n}$ and the set Π of permutations of the integers from 1 to n, find a permutation $\pi=(\pi(1), \pi(2), ..., \pi(n))\in\Pi$ that minimizes

$$z(\pi) = \sum_{i=1}^{n-1} d_{\pi(i),\pi(i+1)} + d_{\pi(n),\pi(1)} \quad (1)$$

The interpretation of n, D and π is as follows: n is the number of cities; D is the matrix of distances between all pairs of these cities; $j = \pi(i)$ denotes city j to visit at step i. Usually, permutations are called tours, and the pairs $(\pi(1), \pi(2))$, ..., $(\pi(i), \pi(i+1))$, ..., $(\pi(n), \pi(1))$ are called edges. So, solving the TSP means searching for the shortest closed tour in which every city is visited exactly once. In other words, our goal is to find an ordering π or tour, of the cities that minimizes the length of the tour, given by Eq. (1).

In this work, we will restrict our attention to the two-dimensional Euclidean TSP, which is the special case where the cities are points in the plane and d_{ij} is the Euclidean distance from city i to city j.

If the distances satisfy $d_{ij} = d_{ji}$ for $1 \le i, j \le N$, this case is the symmetric TSP. However, it is possible to discard that last condition and allow the distance from city i to city j to be different from the distance between city j and city i. We refer to that case as the asymmetric TSP.

3 Differential Evolution Algorithm

A. Classical DE algorithm

DE is a population-based stochastic function to minimize (or maximize) relating to evolutionary algorithms, whose simple yet powerful and straightforward features make it very attractive for numerical optimization.

DE combines simple arithmetical operators with the classical operators of recombination, mutation and selection to evolve from a randomly generated starting population to a final solution. DE uses mutation which is based on the distribution of solutions in the current population. In this way, search directions and possible step sizes depend on the location of the individuals selected to calculate the mutation values [20]. It evolutes generation by generation until the termination conditions have been met.

The different variants of DE are classified using the following notation: DE/α/β/δ, where α indicates the method for selecting the parent chromosome that will form the base of the mutated vector, β indicates the number of difference vectors used to perturb the base chromosome, and δ indicates the recombination mechanism used to create the offspring population. The *bin* acronym indicates that the recombination is controlled by a series of independent binomial experiments.

The fundamental idea behind DE is a scheme whereby it generates the trial parameter vectors. In each step, the DE mutates vectors by adding weighted, random vector differentials to them. If the cost of the trial vector is better than that of the target, the target vector is replaced by the trial vector in the next generation. The variant implemented here was DE/*rand*/1/*bin*, which involved the following steps and procedures:

Step 1: *Initialization of the parameter setup*: The user must choose the key parameters that control DE, i.e., population size, boundary constraints of optimization variables, mutation factor (f_m), crossover rate (*CR*), and the stopping criterion (t_{max}).

Step 2: *Initialize the initial population of individuals*: Initialize the generation's counter, $t = 0$, and also initialize a population of individuals (solution vectors) $x(t)$ with random values generated according to a uniform probability distribution in the n-dimensional problem space.

Step 3: *Evaluate the objective function value*: For each individual, evaluate its objective function (fitness) value.

Step 4: *Mutation operation (or differential operation)*: Mutate individuals according to the following equation:

$$z_i(t+1) = x_{i_1}(t) + f_m \cdot [x_{i_2}(t) - x_{i_3}(t)] \tag{2}$$

where $i = 1,2,\ldots,N$ is the individual's index of population; t is the time (generation); $x_i(t) = \left[x_{i_1}(t), x_{i_2}(t), \ldots, x_{i_n}(t)\right]^{\mathrm{T}}$ stands for the position of the i-th individual of population of N real-valued n-dimensional vectors; $z_i(t) = \left[z_{i_1}(t), z_{i_2}(t), \ldots, z_{i_n}(t)\right]^{\mathrm{T}}$ stands for the position of the i-th individual of a *mutant vector*; $f_m > 0$ is a real parameter, called *mutation factor*, which controls the amplification of the difference between two individuals so as to avoid search stagnation. The mutation operation randomly select the target vector $x_{i_1}(t)$, with $i \neq i_1$.Then, two individuals $x_{i_2}(t)$ and $x_{i_3}(t)$ are randomly selected with $i_1 \neq i_2 \neq i_3 \neq i$, and the difference vector x_{i_2} - x_{i_3} is calculated.

Step 5: *Crossover (recombination) operation*: Following the mutation operation, crossover is applied in the population. For each mutant vector, $z_i(t+1)$, an index

$rnbr(i) \in \{1, 2, \cdots, n\}$ is randomly chosen using a uniform distribution, and a *trial vector*, $u_i(t+1) = [u_{i_1}(t+1), u_{i_2}(t+1), \ldots, u_{i_n}(t+1)]^T$, is generated via

$$u_{i_j}(t+1) = \begin{cases} z_{i_j}(t+1) \text{ if } randb(j) \leq CR \text{ or } j = rnbr(i), \\ x_{i_j}(t) \text{ otherwise}, \end{cases} \quad (3)$$

where j=1,2,..., n is the parameter index; $x_{ij}(t)$ stands for the i-th individual of *j-th* real-valued vector; $z_{ij}(t)$ stands for the i-th individual of *j-th* real-valued vector of a *mutant vector*; $u_{ij}(t)$ stands for the i-th individual of *j-th* real-valued vector after crossover operation; *randb(j)* is the j-th evaluation of a uniform random number generation with [0, 1]; *CR* is a *crossover rate* in the range [0, 1];

To decide whether or not the vector $u_i(t + 1)$ should be a member of the population comprising the next generation, it is compared to the corresponding vector $x_i(t)$. Thus, if f denotes the objective function under minimization, then

$$x_i(t+1) = \begin{cases} u_i(t+1) \text{ if } f(u_i(t+1)) < f(x_i(t)), \\ x_i(t) \text{ otherwise}, \end{cases} \quad (4)$$

Step 6: *Update the generation's counter*: $t = t + 1$;

Step 7: *Verification of the stopping criterion*: Loop to Step 3 until a stopping criterion is met, usually a maximum number of iterations (generations), t_{max}.

B. Discrete DE algorithm

DE algorithms are evolutionary algorithms that have already shown appealing features as efficient methods for the optimization of continuous space functions. The DE algorithms use a floating-point representation for the solutions in the population.

However, the continuous nature of the algorithm prohibits DE to apply to combinatorial optimization problems. To compensate this drawback, Tasgetiren *et al.* [21],[22] presented the Smallest Position Value (SPV) rule, barrowed from the *random key representation* of Bean [23], for the particle swarm optimization (PSO) algorithm, which is developed by Kennedy and Eberhard [24], to convert a continuous position vector to a job permutation. It has been successfully applied to the single machine total weighted tardiness problem and the permutation flow-shop sequencing problem.

The SPV rule can still be used in DE since smallest position value can be replaced by smallest parameter value to convert the continuous parameter values to a permutation. Details about the SPV rule for DE's solutions representation are presented in [21], [22].

For example, with a problem have 6 cities, the first thing is generate a random solution, like 1, 3, 5, 6, 2, 4. After this, it is made a mutation of the continuous values and found a third vector. This vector is changed to a possible solution, e.g. 1, 2, 4, 5, 6, 3. If this solution has less length than the first solution, so, this will be the new best solution. And the process will keep continuing until the stopping criterion is reached.

C. Discrete DE with Local Search

The Lin-Kernighan (LK) heuristic [25] is generally considered to be one of the most effective methods (state-of-the-art of local search heuristics) for generating optimal or near-optimal solutions for the symmetric TSP.

Domain-specific heuristics, such as 2-*opt*, 3-*opt*, and LK, are surprisingly very effective for the TSP. The LK algorithm, also referred to as variable-*opt*, however incorporates a limited amount of hill-climbing by searching for a sequence of exchanges, some of which may individually increase the tour length, but which combine to form a shorter tour. A vast amount has been written about the LK algorithm, including much on its efficient implementation.

Recently, a new and highly effective variant of the LK algorithm has been developed by Helsgaun [16]. This scheme employs a number of important innovations including sequential 5-opt moves and the use of sensitivity analysis to direct the search.

Other alternative approach is the local search based on VNS [17],[18]. Contrary to other heuristics and metaheuristics based on local search methods, VNS does not follow a trajectory but explores increasingly distant neighborhoods of the current incumbent solution, and jumps form this solution to a new one if and only if an improvement has been made.

In this paper, the following discrete DE approaches for the TSP are validated: i) DE approach without local search, ii) DE with local search based on LKH (DE+LKH), iii) DE with local search based on VNS (DE+VNS) and iv) DE+VNS combined with LKH method (DE+VNS+LKH). The figure 1 show the tests realized.

In DE approaches with local search mentioned (DE approaches with sequential hybridization), the DE is used to explore the more promising part of the TSP solution space in order to generate “good” initial solutions, which are refined with LKH and/or VNS.

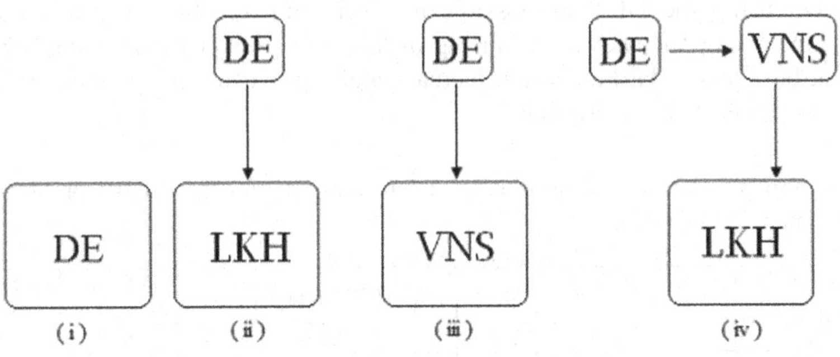

Fig. 1 Representation of the proposed models.

4 Computational Results

Each optimization method was implemented in DEV-C++ development platform with MinGW compiler (ANSI C/C++ compiler) under Windows XP operational system. All the programs were tested in Intel Core Due, 1.6 GHz processor with 1024 MB of random access memory. In each case study, 50 independent runs were made for each of the optimization methods involving 50 different initial trial solutions for each optimization method.

In discrete DE, the population size N was 50 and the stopping criterion t_{max} was 100 generations (5000 evaluations of the objective function). In the LKH and VNS, the stopping criterion t_{max} was 100 generations or if is found best solution. In the VNS, the population size was divided in 6 small groups and it was applied a local search *2-opt*. This local search uses the same stopping criterion of other methods.

The results found with the techniques proposed in this work for the TSPLIB's symmetrical cases are compared with optimal value for each tested benchmark. In this paper, it considers the results of solving the selected four instances in the TSPLIB with the number of cities varying between 101 and 7397. In Tables 1 to 7 are shown the mean errors/deviations of the obtained results in relation to the optimal values of objective function $z(\pi)$ and also a statistical analysis of results for 50 runs.

Looking the results, it is possible to verify that in all the instances the DE was successful in increase the results. All the mean results show a relevant increasing in the performance and the time to find the solution was very fast too.

In the Tables I and II, it is possible to observe the excellent performance of DE + VNS + LKH, DE + LKH, and LKH approaches. In the Table III, it is possible to verify that the standard deviation was better than the original result and the time running the DE+VNS+LKH was faster than the original. Table IV shows that the DE+VNS+LKH obtained 100% of best solution in all runs. And again the result obtained was faster than the original.

Although, the Tables V and VI, the DE+VNS+LKH had an increasing of time, but the standard deviation were less than the others. In neither case the mean reach 100%, but using the DE was possible to verify that the field was get better. Finally, the Table VII shows that in big instances the result found using the DE was fast, but the standard deviation increase and either using or not the VNS, the value was bigger than the original.

Table 1 Results in terms of minimization of function $z(\pi)$ for the EIL101 problem (50 runs, Optimum Value = 629).

Optimization method	Minimum	%	Mean	%	Maximum	%	Standard Deviation	Mean Time (s)
DE	2189	28.73	2276	27.63	2343	26.84	36.11	1.95
DE + VNS	1460	43.08	1612	39.01	1765	35.63	68.35	0.41
DE+VNS+LKH	629	100	629	100	629	100	0.00	0.02
DE + LKH	629	100	629	100	629	100	0.00	< 0.01
LKH	629	100	629	100	629	100	0.00	0.02

Table 2 Results in terms of minimization of function $z(\pi)$ for the KROA150 problem (50 runs, Optimum Value = 26524).

Optimization method	Minimum	%	Mean	%	Maximum	%	Standard Deviation	Mean Time(s)
DE	168126	15.77	177449	14.94	182049	14.56	3008.65	2.58
DE + VNS	107224	24.73	119526	22.19	127406	20.81	3742.95	0.67
DE+VNS+LKH	26524	100	26524	100	26524	100	0.00	0.06
DE + LKH	26524	100	26524	100	26524	100	0.00	0.04
LKH	26524	100	26524	100	26524	100	0.00	0.04

Table 3 Results in terms of minimization of function $z(\pi)$ for the LIN318 problem (50 runs, Optimum Value = 42029).

Optimization method	Minimum	%	Mean	%	Maximum	%	Standard Deviation	Mean Time(s)
DE	467121	8.99	476169	8.82	484940	8.66	4353.08	4.97
DE + VNS	317948	13.21	337404	12.45	359152	11.70	8242.69	1.56
DE+VNS+LKH	42029	100	42066	0.99	42143	0.99	53.51	1.48
DE + LKH	42029	100	42066	0.99	42143	0.99	52.21	0.50
LKH	42029	100	42067	0.99	42143	0.99	52.91	0.50

Table 4 Results in terms of minimization of function $z(\pi)$ for the ATT532 problem (50 runs, Optimum Value = 27686).

Optimization method	Minimum	%	Mean	%	Maximum	%	Standard Deviation	Mean Time (s)
DE	1343428	2.06	1372841	2.01	1395667	1.98	10863.05	8.44
DE + VNS	901324	3.07	965280	2.86	1013743	2.73	24323.70	2.66
DE+VNS+LKH	27686	100	27686	100	27706	0.99	3.95	3.34
DE + LKH	27686	100	27686	100	27703	0.99	2.40	1.18
LKH	27686	100	27688	0.99	27706	0.99	5.71	1.38

Table 5 Results in terms of minimization of function $z(\pi)$ for the U2152 problem (50 runs, Optimum Value = 64253).

Optimization method	Minimum	%	Mean	%	Maximum	%	Standard Deviation	Mean Time (s)
DE	2358326	2.72	2387348	2.69	2400930	2.67	9857.27	39.15
DE + VNS	1865125	3.44	1909209	3.33	1954569	3.28	17958.15	11.33
DE+VNS+LKH	64253	100	64281	0.99	64337	0.99	23.15	237.38
DE + LKH	64253	100	64272	0.99	64324	0.99	22.60	83.08
LKH	64253	100	64278	0.99	64324	0.99	23.57	86.58

Table 6 Results in terms of minimization of function $z(\pi)$ for the PCB3038 problem (50 runs, Optimum Value = 137694).

Optimization method	Minimum	%	Mean	%	Maximum	%	Standard Deviation	Mean Time (s)
DE	5133870	2.68	5170664	2.66	5202107	2.64	16979.24	59.10
DE + VNS	4251942	3.23	4302541	3.20	4356353	3.16	26434.48	17.06
DE+VNS+LKH	137694	100	137704	0.99	137753	0.99	20.78	369.34
DE + LKH	137694	100	137706	0.99	137753	0.99	23.34	123.10
LKH	137694	100	137710	0.99	137757	0.99	22.55	142.64

Table 7 Results in terms of minimization of function $z(\pi)$ for the PLA7397 problem (50 runs, Optimum Value = 23260728).

Optimization method	Minimum	%	Mean	%	Maximum	%	Standard Deviation	Mean Time (s)
DE	2693563800	0.86	2710405086	0.85	2723186572	0.85	6252729.91	193.23
DE + VNS	2282677680	1.01	2292893726	1.01	2305226654	1.00	7921297.27	131.94
DE+VNS+LKH	23260728	100	23261170	0.99	23264711	0.99	1327.66	13636.55
DE + LKH	23260728	100	23261533	0.99	23265152	0.99	1346.24	6771.52
LKH	23260728	100	23261000	0.99	23264711	0.99	1294.50	7441.50

5 Conclusion

In this paper, hybrid discrete DE approaches with local search based on VNS and/or LKH were described and evaluated the quality of solutions on instances of TSPLIB, and experiments showed its validity (see Tables 1-7).

As it is possible to verify in the Tables I to IV, the result obtained only using the DE was not enough to find a final solution for the TSP problem. But, after applied the result found for an initial solution in the LKH problem, it is clear that obtained an increasing in the performance.

It is possible to analyze that for small instances the code is not worth to use, but in big instances the results obtained were good and the time to run fall down to 15% of the original time running only LKH.

In all the tables, all the configuration parameters were the same, but this could be modified to increase more the results. One parameter that could be modified is the DE method used. In this paper, it was used the method 1 (DE/*rand*/1/*bin*), but there are more 9 possible methods based on [10],[11].

Moreover, the computational results of presented hybrid discrete DE approaches with VNS and/or LKH are very close to the best-known TSPLIB solution values. Effective implementation of these and related neighborhoods in discrete DE approaches are topics for further investigation in multi-objective TSPs.

Acknowledgment

The authors would like to thank National Council of Scientific and Technologic Development of Brazil - CNPq (projects: 568221/2008-7, 474408/2008-6,

302786/2008-2-PQ, 303963/2009-3/PQ and 478158/2009-3) and 'Fundação Araucária' (project: 14/2008-416/09-15149) for its financial support of this work.

References

[1] Aarts, E., Lenstra, J.K. (eds.): Local Search in Combinatorial Optimization. Princeton University Press, Princeton (2003)

[2] Colorni, A., Dorigo, M., Maffioli, F., Maniezzo, V., Righini, G., Trubian, M.: Heuristics from nature for hard combinatorial optimization problems. International Transactions in Operational Research 3(1), 1–21 (1996)

[3] Laporte, G.: The traveling salesman problem: An overview of exact and approximate algorithms. European Journal of Operational Research 59(2), 231–247 (1992)

[4] Bektas, T.: The multiple traveling salesman problem: an overview of formulations and solution procedures. Omega 34(3), 209–219 (2006)

[5] Lawler, E.L., Lenstra, J.K., Kan, A.H.G.R., Shmoys, D.B. (eds.): The Traveling Salesman Problem: A Guided Tour of Combinatorial Optimization. Wiley, New York (1985)

[6] Ding, Z., Leung, H., Zhu, Z.: A study of the transiently chaotic neural network for combinatorial optimization. Mathematical and Computer Modelling 36(9-10), 1007–1020 (2002)

[7] Katayama, K., Sakamoto, H., Narihisa, H.: The efficiency of hybrid mutation genetic algorithm for the travelling salesman problem. Mathematical and Computer Modelling 31(10-12), 197–203 (2000)

[8] Carter, A.E., Ragsdale, C.T.: A new approach to solving the multiple traveling salesperson problem using genetic algorithms. European Journal of Operational Research 175(1), 246–257 (2006)

[9] Storn, R., Price, K.: Differential evolution: a simple and efficient adaptive scheme for global optimization over continuous spaces. Technical Report TR-95-012, International Computer Science Institute, Berkeley, CA, USA (1995)

[10] Storn, R., Price, K.: Differential evolution a simple and efficient heuristic for global optimization over continuous spaces. Journal of Global Optimization 11(4), 341–359 (1997)

[11] Tasgetiren, M.F., Pan, Q.K., Liang, Y.C., Suganthan, P.N.: A discrete differential evolution algorithm for the total earliness and tardiness penalties with a common due date on a single-machine. In: Proceedings of IEEE Symposium on Computational Intelligence in Scheduling, Honolulu, HI, USA, pp. 271–278 (2007)

[12] Tasgetiren, M.F., Pan, Q.K., Suganthan, P.N., Liang, Y.C.: A discrete differential evolution algorithm for the no-wait flowshop scheduling problem with total flowtime criterion. In: Proceedings of IEEE Symposium on Computational Intelligence in Scheduling, Honolulu, HI, USA, pp. 251–258 (2007)

[13] Tasgetiren, M.F., Suganthan, P.N., Pan, Q.K.: A discrete differential evolution algorithm for the permutation flowshop scheduling problem. In: Proceedings of Genetic and Evolutionary Computation Conference, London, UK, pp. 158–167 (2007)

[14] Qian, B., Wang, L., Huang, D.-X., Wang, W.-L., Wang, X.: An effective hybrid DE-based algorithm for multi-objective flow shop scheduling with limited buffers. Computers & Operations Research 36(1), 209–233 (2009)

[15] Onwubolu, G., Davendra, D.: Scheduling flow shops using differential evolution algorithm. European Journal of Operational Research 171(2), 674–692 (2006)
[16] Helsgaun, K.: An effective implementation of the Lin–Kernighan traveling salesman heuristic. European Journal of Operational Research 126(1), 106–130 (2000)
[17] Mladenovic, N.: A variable neighborhood algorithm – a new metaheuristic for combinatorial optimization. In: Abstracts of Papers at Optimization Days, Montreal, Canada, vol. 112 (1995)
[18] Mladenovic, N., Hansen, P.: Variable neighbourhood search. Computers and Operations Research 24(11), 1097–1100 (1997)
[19] Reinelt, G.: TSPLIB – a traveling salesman problem library. ORSA Journal on Computing 4, 134–143 (1996), http://www.iwr.uni-heidelberg.de/groups/comopt/software/
[20] Montes, E.M., Reyes, J.V., Coello, C.A.C.: A comparative study of differential evolution variants for global optimization. In: Proceedings of Genetic and Evolutionary Computation Conference, Seattle, Washington, USA (2006)
[21] Tasgetiren, M.F., Sevkli, M., Liang, Y.-C., Gencyilmaz, G.: Particle swarm optimization algorithm for single-machine total weighted tardiness problem. In: Proceedings of Congress on Evolutionary Computation, Portland, Oregon, USA, vol. 2, pp. 1412–1419 (2004)
[22] Tasgetiren, M.F., Sevkli, M., Liang, Y.-C., Gencyilmaz, G.: Particle swarm optimization algorithm for permutation flowshop sequencing problem. In: Dorigo, M., Birattari, M., Blum, C., Gambardella, L.M., Mondada, F., Stützle, T. (eds.) ANTS 2004. LNCS, vol. 3172, pp. 382–389. Springer, Heidelberg (2004)
[23] Bean, J.C.: Genetic algorithm and random keys for sequencing and optimization. ORSA Journal on Computing 6(2), 154–160 (1994)
[24] Eberhard, R.C., Kennedy, J.: A new optimizer using particle swarm theory. In: Proceedings of the Sixth International Symposium on Micro Machine and Human Science, Nagoya, Japan, pp. 39–43 (1995)
[25] Lin, S., Kernighan, B.W.: An effective heuristic algorithm for the traveling salesman problem. Operations Research 21, 498–516 (1973)

Genetic Algorithm Based Reliability Optimization in Interval Environment

A.K. Bhunia[1] and L. Sahoo[2]

[1] Department of Mathematics, The University of Burdwan, Burdwan-713104, West Bengal, India
math-akbhunia@buruniv.ac.in
[2] Department of Mathematics, Raniganj Girls College, Raniganj-713347, West Bengal, India
lxsahoo@gmail.com

The objective of this chapter is to develop and solve the reliability optimization problems of series-parallel, parallel-series and complicated system considering the reliability of each component as interval valued number. For optimization of system reliability and system cost separately under resource constraints, the corresponding problems have been formulated as constrained integer/mixed integer programming problems with interval objectives with the help of interval arithmetic and interval order relations. Then the problems have been converted into unconstrained optimization problems by two different penalty function techniques. To solve these problems, two different real coded genetic algorithms (GAs) for interval valued fitness function with tournament selection, whole arithmetical crossover and non-uniform mutation for floating point variables, uniform crossover and uniform mutation for integer variables and elitism with size one have been developed. To illustrate the models, some numerical examples have been solved and the results have been compared. As a special case, taking lower and upper bounds of the interval valued reliabilities of component as same the corresponding problems have been solved and the results have been compared with the results available in the existing literature. Finally, to study the stability of the proposed GAs with respect to the different GA parameters (like, population size, crossover and mutation rates), sensitivity analyses have been shown graphically.

1 Introduction

While advanced technologies have raised the world to an unprecedented level of productivity, our modern society has become more delicate and vulnerable due to the increasing dependence on modern technological systems that

N. Nedjah et al. (Eds.): Innovative Computing Methods, SCI 357, pp. 13–36.
springerlink.com

often require complicated operations and highly sophisticated management. From any respect, the system reliability is a crucial measure to be considered in systems operation and risk management. When designing a highly reliable system, there arises an important question as to how to obtain a balance between reliability and other resources e.g., cost, volume and weight. In the last few decades, several researchers considered reliability optimization problems, like redundancy allocation and cost minimization problems as integer nonlinear programming problems (INLPP) and/or mixed-integer nonlinear programming problems (MINLPP) with single or several resource constraints [1-14]. To solve those problems, different techniques have been proposed by the several researchers. In their works, the reliability of each component is known and fixed positive number which lies between zero and one. However, in real life situations, the reliability of an individual component may not be fixed. It may vary due to several reasons. There is no technology by which different components can be produced with exactly identical reliabilities. So, the reliability of each component is sensible and it may be treated as a positive imprecise number instead of a fixed real number. Studies of the system reliability where the component reliabilities are imprecise and/or interval valued have already been initiated by some authors [15-19].To tackle the problem with such imprecise numbers, generally stochastic, fuzzy and fuzzy- stochastic approaches are applied and the corresponding problems are converted to deterministic problems for solving them. In the stochastic approach, the parameters are assumed to be random variables with known probability distributions. In the fuzzy approach, the parameters, constraints and goals are considered as fuzzy sets with known membership functions or fuzzy numbers. On the other hand, in the fuzzy-stochastic approach, some parameters are viewed as fuzzy sets/fuzzy numbers and others as random variables. However, it is a formidable task for a decision maker to specify the appropriate membership function for fuzzy approach and probability distribution for stochastic approach and both for fuzzy -stochastic approach. So, to avoid these difficulties for handling the imprecise numbers by different approaches, one may use intervals number to represent an imprecise number, as this representation is the most significant representation among others. Due to this representation, the system reliability would be interval valued. Here, we have considered GA-based approaches for solving reliability optimization problems with the interval objective. As the objective function of the reliability optimization is interval valued, to solve this type of problem by the GA method, order relations of interval numbers are essential for selection operation as well as for finding the best chromosome in each generation. Here we consider the definition of order relations developed by Mahato and Bhunia [20] in the context of the optimistic and pessimistic decision maker's point of view for maximization and minimization problems.

In this chapter, we have considered the problem of constrained redundancy allocation in the series system, the hierarchical series-parallel system, the complicated or non-parallel-series system and the network reliability system with interval valued reliability components (redundancy allocation and network cost minimization). The problems are formulated as non-linear constrained integer programming problems and/or mixed integer programming problems with interval coefficients [21-22] for maximizing the overall system reliability and system cost under some resource/budget constraints. During the last few years, several techniques were proposed for solving the constrained optimization problem with fixed coefficients with the help of GAs [23-29]. Among these methods, penalty function techniques are very popular in solving the same by GAs [30-32]. This method transforms the constrained optimization problem to an unconstrained optimization problem by penalizing the objective function corresponding to the infeasible solution. Hence, to solve the constrained optimization problem the problem is converted to unconstrained one by two different types of penalty techniques and the resulting objective function would be interval valued. So, to solve this problem we have developed two different GAs for integer variables with the same GA operators like tournament selection, uniform crossover for integer variables and whole arithmetical crossover for floating point variables, uniform mutation for integer variables and boundary mutation for floating point variables and elitism of size one but different fitness function depending on different penalty approaches. These methods have been illustrated with some numerical examples and to test the performance of these methods, results have also been compared. As a special case considering the lower and upper bounds of interval valued reliabilities of components as same, the resulting problem becomes identical with the existing problem available in the literature.

2 Finite Interval Arithmetic

An interval number is a closed interval denoted by $A = [a_L, a_R]$ and is defined by $A = [a_L, a_R] = \{x : a_L \leq x \leq a_R, x \in \Re\}$ where a_L and a_R are the left and right limits respectively and $\Re$ is the set of all real numbers. A can also be expressed in terms of centre and radius as $A = \langle a_c, a_w \rangle = \{x : a_c - a_w \leq x \leq a_c + a_w, x \in \Re\}$, where a_c anda_w are the centre and radius of the interval A respectively, i.e., $a_c = (a_L + a_R)/2$, and $a_w = (a_R - a_L)/2$. Actually, every real number can be treated as an interval, such as for all $x \in \Re$, x can be written as an interval $[x, x]$ having zero width. Now we shall present the definitions of arithmetical operations like addition, subtraction, multiplication, division and integral power of interval numbers [33] and also the n-th root as well as the rational powers of interval numbers [34].

Definition 1: Let $A = [a_L, a_R]$ and $B = [b_L, b_R]$ be two intervals. Then the definitions of addition, scalar multiplication, subtraction, multiplication and division of interval numbers are as follows:

- **Addition:** $A+B=[a_L,a_R]+[b_L,b_R]=[a_L+b_L,a_R+b_R]$
- **Scalar multiplication:** For any real number α,$\alpha A = \alpha[a_L,a_R] = \begin{cases}[\alpha a_L,\alpha a_R] \text{ if } \alpha \geq 0\\ [\alpha a_R,\alpha a_L] \text{ if } \alpha < 0\end{cases}$
- **Subtraction:** $A-B=[a_L,a_R]-[b_L,b_R]=[a_L,a_R]+[-b_R,-b_L]=[a_L-b_R,a_R-b_L]$
- **Multiplication:** $A\times B=[a_L,a_R]\times[b_L,b_R]$ $=[\min(a_Lb_L,a_Lb_R,a_Rb_L,a_Rb_R),\max(a_Lb_L,a_Lb_R,a_Rb_L,a_Rb_R)]$
- **Division** $\frac{A}{B}=A\times\frac{1}{B}=[a_L,a_R]\times[\frac{1}{b_R},\frac{1}{b_L}]$, *provided* $0\notin[b_L,b_R]$

Definition 2: Let $A=[a_L,a_R]$ be an interval and n be any non-negative integer, then

$$A^n=\begin{cases}[1,1] \text{ if } n=0\\ [a_L^n,a_R^n] \text{ if } a_L\geq 0 \text{ or if } n \text{ is odd}\\ [a_R^n,a_L^n] \text{ if } a_R=0 \text{ and } n \text{ is even}\\ [0,\max(a_L^n,a_R^n)] \text{ if } a_L\leq 0\leq a_R \text{ and } n(>0) \text{ is even}\end{cases}$$

Definition 3: The $n-th$ root of an interval $A=[a_L,a_R]$, n being a positive integer, is defined as

$$(A)^{\frac{1}{n}}=[a_L,a_R]^{\frac{1}{n}}=\begin{cases}\sqrt[n]{[a_L,a_R]}=[\sqrt[n]{a_L},\sqrt[n]{a_R}] \text{ if } a_L\geq 0 \text{ or if } n \text{ is odd}\\ [0,\sqrt[n]{a_R}] \quad \text{if } a_L\leq 0, a_R\geq 0 \text{ and } n \text{ is even}\\ \phi \text{ if } a_R<0 \text{ and } n \text{ is even}\end{cases}$$

where ϕ is the empty interval.

Again, by applying the definitions of power and different roots of an interval, we can find any rational power of an interval. For example $A^{\frac{p}{q}}$ obtained by defining $A^{\frac{p}{q}}$ as $(A^p)^{\frac{1}{q}}$.

3 Order Relation of Interval Numbers

Further, for arriving at the optimum solution involving interval algebra, we need to define the order relation of interval numbers.

Let $A=[a_L,a_R]$ and $B=[b_L,b_R]$ be two intervals. These two intervals may be one of the following three types:

1. Type-1: Two intervals are disjoint [see Fig.1].
2. Type-2: Two intervals are partially overlapping [see Fig.2].
3. Type-3: One of the intervals contains the other one [see Fig.3].

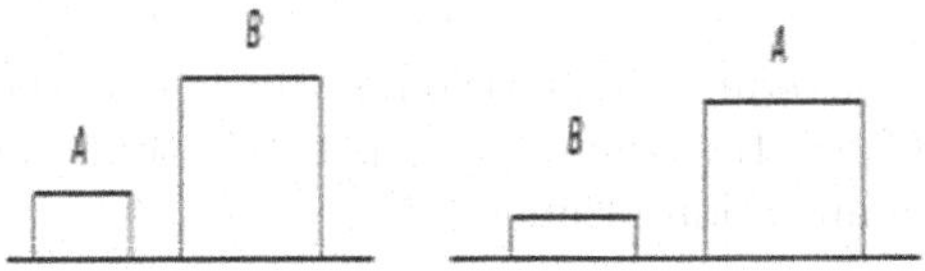

Fig. 1 Type-1 interval

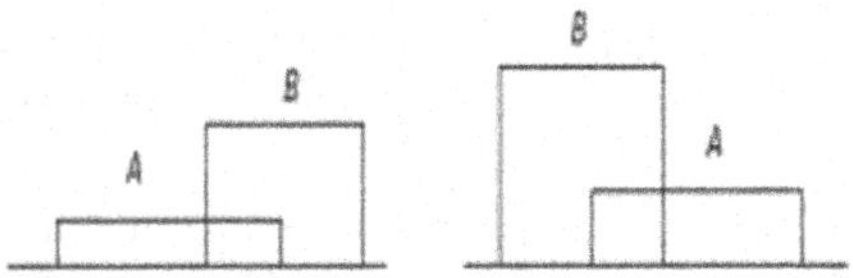

Fig. 2 Type-2 intervals

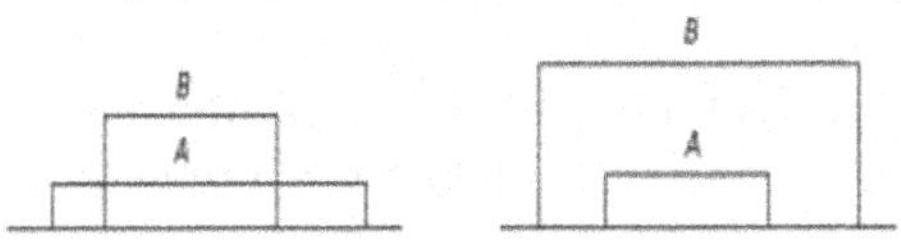

Fig. 3 Type-3 intervals

Here we consider the definitions of order relations developed by Mahato and Bhunia [20] in the context of optimistic and pessimistic decision makers' point of view.

3.1 Optimistic Decision-Making

In optimistic decision-making, decision maker prefers the lowest value for minimization problems and highest value for maximization problems ignoring the uncertainty.

Definition 4: Let us define the order relation $\geq_{\text{omax}}$ between the intervals $A = [a_L, a_R]$ and $B = [b_L, b_R]$ then for maximization problems $A \geq_{\text{omax}} B$ $\Leftrightarrow a_R > b_R$,$A >_{\text{omax}} B \Leftrightarrow A \geq_{\text{omax}} B \wedge A \neq B$.

According to this definition, the optimistic decision maker accepts A . The order relation $\geq_{\text{omax}}$ is reflexive and transitive but not symmetric.

Definition 5: The order relation $\leq_{\text{omin}}$ between the intervals $A = [a_L, a_R]$ and $B = [b_L, b_R]$ then for minimization problems $A \leq_{\text{omin}} B \Leftrightarrow a_L \leq b_L$, $A <_{\text{omin}} B \Leftrightarrow A \leq_{\text{omin}} B \wedge A \neq B$. The order relation $\leq_{\text{omin}}$ is not symmetric.

3.2 Pessimistic Decision-Making

In pessimistic decision making, the decision maker prefers the highest/lowest value under the principle "Less uncertainty is better than more uncertainty" for maximization/minimization problems.

Definition 6: The order relation $>_{\text{pmax}}$ between the intervals $A = [a_L, a_R] = \langle a_c, a_w \rangle$ and $B = [b_L, b_R] = \langle b_c, b_w \rangle$, then for maximization problems

$$\text{(i)} A >_{\text{pmax}} B \Leftrightarrow a_c > b_c \;\; \text{for type} - 1 \,\text{and type} - 2 \text{ intervals,}$$
$$\text{(ii)} A >_{\text{pmax}} B \Leftrightarrow a_c \geq b_c \wedge a_w < b_w \;\;\; \text{for type} - 3 \text{ intervals}$$

However, for Type-3 intervals, pessimistic decision cannot be taken when $a_c > b_c \wedge a_w > b_w$. In this case, we consider the optimistic decision.

Definition 7: The order relation $<_{\text{pmin}}$ between the intervals $A = [a_L, a_R] = \langle a_c, a_w \rangle$ and $B = [b_L, b_R] = \langle b_c, b_w \rangle$, then for minimization problems

$$\text{(i)} A <_{\text{pmin}} B \Leftrightarrow a_c < b_c \;\; \text{for type} - 1\text{and type}{-}2 \text{ intervals,}$$
$$\text{(ii)} A <_{\text{pmin}} B \Leftrightarrow a_c \leq b_c \wedge a_w < b_w \;\;\; \text{for type} - 3 \text{ intervals}$$

However, for Type-3 intervals, pessimistic decision cannot be taken when $a_c < b_c \wedge a_w > b_w$. In this case, we consider the optimistic decision.

4 Assumptions and Notations

Witout loss of generality, tet us assume the following:

- The component reliabilities are imprecise and interval valued.
- The failure of any component is independent of that of the other components.
- All redundancy is active redundancy without repair.

The following notations have been used in the entire paper.

- x_j: the number of redundant components in j-th subsystem
- r_j: reliability of j-th component
- $R_j(x)$: $1-(1-r_j)^{x_j}, j = 1, 2, ..., q$, the reliability of j-th parallel subsystem
- x: $(x_1, x_2, ..., x_n)$
- r_{jL},r_{jR}: lower and upper limits of r_j
- m: number of resource constraints
- n: number of stages of the system
- $R_{jL}(x)$: lower bound of $R_j(x)$
- $R_{jR}(x)$: upper bound of $R_j(x)$
- Q_j: $1 - R_j$
- R_j: the reliability of j-th subsystem, $j = q + 1, q + 2, \cdots, n$

- (x, R): $(x_1, x_2, ..., x_q, R_{q+1}, ..., R_n)$
- $R_S(x, R)$: system reliability
- $R_{SL}(x, R)$: lower bound of $R_S(x, R)$
- $R_{SR}(x, R)$: upper bound of $R_S(x, R)$
- $C_i(x, R)$: consumption of i-th resource $(i = 1, 2, ..., m)$
- $C_w(x, R)$: weighted cost
- c_i: availability of i-th resource $(i = 1, 2, ..., m)$
- l_j,u_j: lower and upper bounds of x_j
- α_j,β_j: lower and upper bounds of R_j, $j = q + 1, q + 2, \cdots, n$
- R^*: minimum prescribed reliability in case of cost minimization problem
- *p_size*: population size
- *p_cross*: probability of crossover or crossover rate
- *p_mute*: probability of mutation or mutation rate

5 Constrained Redundancy Optimization Problem for different Systems

5.1 Series System

It is well known that a series system (ref. Fig. 4) with n independent components must be operating only if all the components are functioning. In order to improve the overall reliability of the system; one can use more reliable components. However, the expenditure and more often the technological limits may prohibit an adoption of this strategy. An alternative technique is to add some redundant components as shown in Fig. 5. The goal of the problem is to determine an optimal redundancy allocation so as to maximize the overall system reliability under limited resource constraints. These constraints may arise out of the size, cost and quantities of the resources. Mathematically, the constrained redundancy optimization problem for such a system for interval valued of reliability can be formulated as follows:

Problem-1: Maximize $[R_{SL}, R_{SR}] = \prod_{j=1}^{q} [\{1 - (1 - r_{jL})^{x_j}\}, \{1 - (1 - r_{jR})^{x_j}\}]$

subject to $g_i(x) \leq c_i$, $i = 1, 2, ..., m$ and $l_j \leq x_j \leq u_j$, for $j = 1, 2, ..., q$, where $r_j = [r_{jL}, r_{jR}]$

This is a constrained nonlinear integer programming problem with interval valued objective.

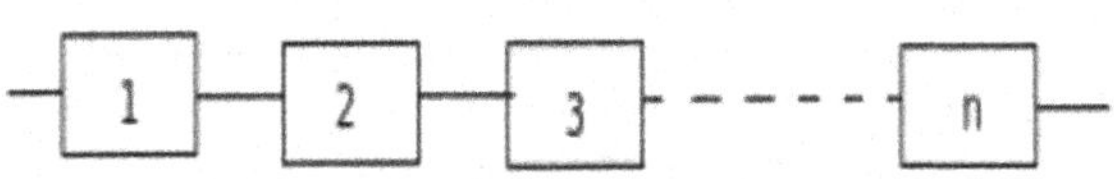

Fig. 4 Series System

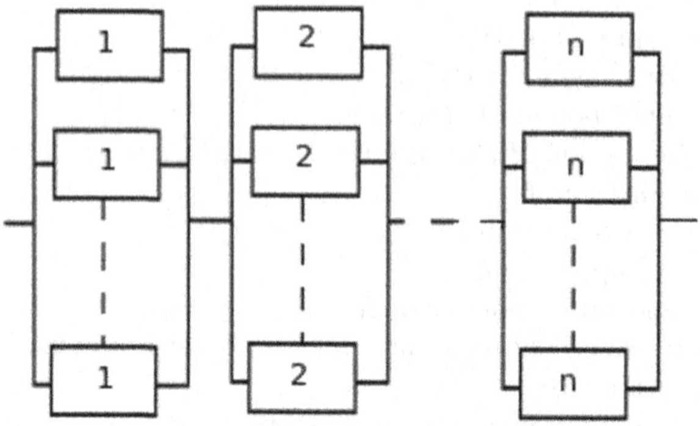

Fig. 5 Parallel series system

5.2 Hierarchical Series-Parallel System

A reliability system is called a hierarchical series parallel system (HSP) if the system can be viewed as a set of subsystems arranged in a series parallel; each subsystem has a similar configuration; subsystems of each subsystem have a similar configuration and so on. For example, let us consider a HSP system ($n = 10, m = 2$) shown in the Fig.6. This system has a nonlinear and non separable structure and consists of nested parallel and series system. The system reliability of HSP is given by $R_S = \{1 - \langle 1 - [1 - Q_3(1 - R_1R_2)]R_4\rangle(1 - R_5R_6)\}(1 - Q_7Q_8Q_9)R_{10}$. Mathematically; the constrained redundancy optimization problem for this system for interval valued reliability can be formulated as follows:

Problem-2: Maximize $[R_{SL}, R_{SR}] = \{1-\langle 1-(1-[Q_{3L}, Q_{3R}](1-[R_{1L}, R_{1R}]$ $[R_{2L}, R_{2R}]))\ [R_{4L}, R_{4R}]\rangle(1-[R_{5L}, R_{5R}][R_{6L}, R_{6R}])\}(1-[Q_{7L}, Q_{7R}][Q_{8L}, Q_{8R}]$

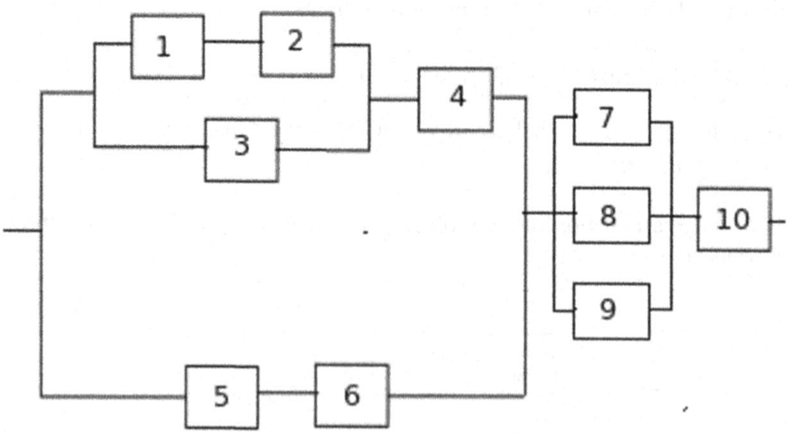

Fig. 6 Hierarchical series-parallel system

$[Q_{9L}, Q_{9R}])$ $[R_{10L}, R_{10R}]$ subject to $g_i(x) \leq c_i, \quad i = 1, 2, ..., m$ and $l_j \leq x_j \leq u_j$ for $j = 1, 2, ..., q$. This is an INLP with interval valued objective.

5.3 Complicated System

When a reliability system can be reduced to series and parallel configurations, there exist combinations of components which are connected neither in a series nor in parallel. Such systems are called complicated or non parallel series systems. This system is also called the bridge system. For example, let us consider a bridge system ($n = 5, m = 3$) shown in Fig.7. This system consists of five subsystems and three nonlinear and non-separable constraints. The overall system reliability R_S is given by the expression as follows:

$$R_S = R_5(1 - Q_1Q_3)(1 - Q_2Q_4) + Q_5[1 - (1 - R_1R_2)(1 - R_3R_4)]$$

$$R_i = R_i(x_i) \text{ and } Q_i = 1 - R_i \text{ for all } i = 1, 2, 3, 4, 5 \; .$$

Mathematically, the constrained redundancy optimization problem for such complicated system for interval valued reliability can be formulated as follows:

Problem-3: Maximize $[R_{SL}, R_{SR}] = [R_{5L}, R_{5R}](1 - [Q_{1L}, Q_{1R}][Q_{3L}, Q_{3R}])$ $(1 - [Q_{2L}, Q_{2R}][Q_{4L}, Q_{4R}]) + [Q_{5L}, Q_{5R}]\{1 - (1 - [R_{1L}, R_{1R}][R_{2L}, R_{2R}])(1 - [R_{3L}, R_{3R}][R_{4L}, R_{4R}])\}$ subject to $g_i(x) \leq c_i, \quad i = 1, 2, ..., m$ and $l_j \leq x_j \leq u_j$, for $j = 1, 2, ..., q$

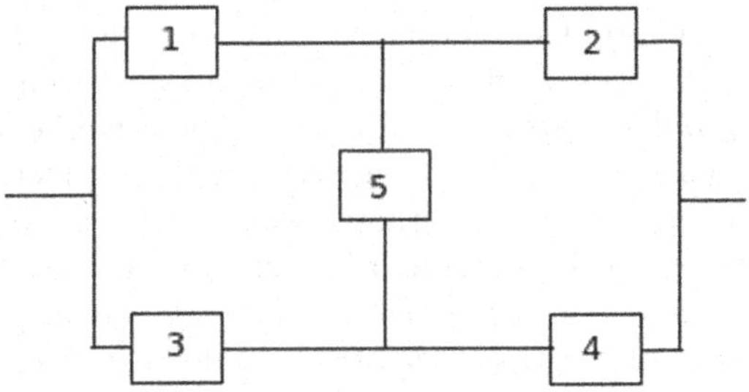

Fig. 7 Complicated system

5.4 k-out-of-n System

A $k - out - of - n$ system is a n-component system which functions when at least k of its components function. This redundant system is sometimes used in the place of pure parallel system. It is also referred to as $k - out - of - n : G$ system. An n-component series system is a $n - out - of - n : G$ system whereas a parallel system with n-components is a $1 - out - of - n : G$ system. When all of the components are independent and identical, the reliability of

$k - out - of - n$ system can be written as $R_S = \sum_{j=k}^{n} \binom{n}{j} r^j (1-r)^{n-j}$, where r is the component reliability.

5.5 *Reliability Network System*

Let us consider a network with n subsystems. The goal of the redundancy allocation problem is to determine the number of redundant components in each of q parallel subsystems and reliability levels of $(n-q)$ general subsystems so as to maximize the overall system reliability subject to the given resource constraints and also to minimize the overall system cost subject to the given constraint on system reliability. The corresponding problems are mixed-integer nonlinear programming problems as follows:

Problem-4: Maximize $R_S(x,R) = f(R_1(x_1), R_2(x_2), ..., R_q(x_q), R_{q+1}, ..., R_n)$ subject to $C_i(x,R) \le c_i,\ i = 1, ..., m,$ and $1 \le l_j \le x_j \le u_j,\ x_j$ integer, $j = 1, ..., q,\ 0 < \alpha_j \le R_{j+1} \le \beta_j < 1,\ \ j = 1, ..., n-q,$

Problem-5: Minimize$C_w(x,R)$ subject to $R_S(x,R) \ge R^*$, where $R_S(x,R) = f(R_1(x_1), R_2(x_2), ..., R_q(x_q), R_{q+1}, ..., R_n)$

6 GA Based Constrained Handling Technique

In the application of GA for solving reliability optimization problem with interval objective, there arises an important question for handling the constraints relating to the problem. During the past, several methods have been proposed to handle the constraints in evolutionary algorithms [30], [32] for solving the same problem with fixed objective. These methods can be classified into several types, viz. penalty function techniques, methods that preserve the feasibility of solutions, methods that clearly distinguish between feasible and infeasible solutions and hybrid methods. Among these methods, penalty function technique is very well known and widely applicable. In this technique, the amount of constraint violations is added /subtracted to the objective function in different ways. When the objective function is increased/ decreased with a penalty term multiplied by so called penalty coefficient there arises a difficulty to select the initial value and upgrading strategy for the penalty coefficient. To overcome this difficulty, Deb [30] proposed a GA based Parameter Free Penalty (PFP) technique. In this technique, the worst fitness value of GA for feasible solutions is considered as the fitness value of infeasible solution without multiplying the penalty coefficient i.e., the fitness function values of infeasible solutions are independent of the objective function value for the same solution. Therefore, according to the PFP technique, the converted problem of problem (1-3) is as follows:

$$\text{Maximize } [\hat{R}_{SL}(x), \hat{R}_{SR}(x)] = [R_{SL}(x), R_{SR}(x)] - [\sum_{i=1}^{m} \max(0, g_i(x) - c_i), \sum_{i=1}^{m} \max(0, g_i(x) - c_i)] + \theta(x) \quad (1)$$

where $\theta(x) = \begin{cases} [0,0] & \text{if } x \in S \\ -[R_{SL}(x), R_{SR}(x)] + \min[R_{SL}, R_{SR}] & \text{if } x \notin S \end{cases}$
and $S = \{x : g_i(x) \leq c_i, i = 1, 2, ..., m \text{ and } l \leq x \leq u\}$

Here $\min[R_{SL}, R_{SR}]$ is the value of interval valued objective function of the worst feasible solution in the population. Alternatively, the problem may be solved with another fitness function by penalizing a large positive number (say M which can be written in the interval form as$[M, M]$) [18]. This penalty function method is known as Big-M penalty and its form is as follows:

$$\text{Maximize } [\hat{R}_{SL}(x), \hat{R}_{SR}(x)] = [R_{SL}(x), R_{SR}(x)] + \theta(x) \quad (2)$$

where $\theta(x) = \begin{cases} [0,0] \text{ if } x \in S \\ -[R_{SL}(x), R_{SR}(x)] + [-M, -M] \text{ if} x \notin S \end{cases}$
and $S = \{x : g_i(x) \leq c_i, i = 1, 2, ..., m \text{ and } l \leq x \leq u\}$

The above problems (1) and (2) are nonlinear unconstrained integer programming problem with interval coefficients. Also, according to the PFP technique, the converted problem of problem-4 is as follows:

$$\text{Maximize } \hat{R}_S(x, R) = R_S(x, R) - \sum_{i=1}^{m} [\max(0, C_i(x, R) - c_i), \max(0, C_i(x, R) - c_i)] + \theta(x, R) \quad (3)$$

where $\theta(x, R) = \begin{cases} [0,0] & , \quad if\ (x, R) \in X \\ -R_S(x, R) + \min[R_{SL}(x, R), R_{SR}(x, R)] & if\ (x, R) \notin X \end{cases}$
and $X = \{(x, R) : C_i(x, R) \leq c_i,\ i = 1, ..., m \text{ and } l \leq x \leq u, \alpha \leq R \leq \beta\}$

Here $\min[R_{SL}(x, R), R_{SR}(x, R)]$ is the value of the interval valued objective function of the worst feasible solution in the population.

Alternatively, the problem may also be solved with another fitness function by penalizing a large positive number. The converted form is as follows:

$$\text{Maximize } \hat{R}_S(x, R) = R_S(x, R) + \theta(x, R) \quad (4)$$

where $\theta(x, R) = \begin{cases} [0,0] & if\ (x, R) \in X \\ -R_S(x, R) + [-M, -M] & if\ (x, R) \notin X \end{cases}$
and $X = \{(x, R) : C_i(x, R) \leq c_i,\ i = 1, ..., m \text{ and } l \leq x \leq u, \alpha \leq R \leq \beta\}$

Similarly, for Problem-5, the converted problem is as follows:

$$\text{Minimize } \hat{C}_w(x,R) = C_w(x,R) + \sum_{j=1}^{m}[\max(0, -R_{SL}(x,R) + R^*)] + \theta(x,R) \tag{5}$$

where $\theta(x,R) = \begin{cases} [0,0] & if\,(x,R) \in X \\ -C_w(x,R) + \max\{C_w(x,R)\} & if\,(x,R) \notin X \end{cases}$
and $X = \{(x,R) : -R_{SL}(x,R) + R^* \leq 0,\ i = 1,2,...,m \text{ and } l \leq x \leq u, \alpha \leq R \leq \beta\}$

Here $\max\{C_w(x,R)\}$ is the value of the interval valued objective function of the worst feasible solution in the population. Alternatively, the problem may also be solved with another fitness function by penalizing a large positive number. The converted problem is of the form

$$\text{Minimize } \hat{C}_w(x,R) = C_w(x,R) + \theta(x,R) \tag{6}$$

where $\theta(x,R) = \begin{cases} [0,0] & if(x,R) \in X \\ -C_w(x,R) + M & if(x,R) \notin X \end{cases}$
and
$X = \{(x,R) : -R_{SL}(x,R) + R^* \leq 0,\ i = 1,...,m \text{ and } l \leq x \leq u, \alpha \leq R \leq \beta\}$

The above problems (1-2) are non-linear unconstrained integer programming problem with interval coefficients whereas problems (3-6) are non-linear unconstrained mixed integer programming problem with interval coefficients.

7 Genetic Algorithm

Genetic Algorithm is a well-known stochastic method of global optimization based on the evolutionary theory of Darwin: ' The survival of the fittest' and natural genetics (Goldberg [23]). It has successfully been applied in different real world application problems. This algorithm starts with an initial population of chromosomes. These populations will be improved from generation to generation by an artificial evolution process. During each generation, each chromosome in the entire population is evaluated using the measure of fitness and the population of the next generation is created through different genetic operators. This algorithm can be implemented easily with the help of computer programming. In particular, it is very useful for solving complicated optimization problems which cannot be solved easily by analytical /direct/gradient based mathematical techniques.

For implementing the GA in solving the optimization problems, the following basic components are to be considered.

- GA Parameters
- Chromosome representation
- Initialization of population
- Evaluation of fitness function
- Selection process
- Genetic operators (crossover,mutation and elitism)
- Termination criteria

Initially, the chromosomes/individuals are generated randomly. In this work, each chromosome/individual has n components/genes of which first q genes are relating to integer variables whereas the last $(n-q)$ are relating to floating point variable. These chromosomes/individuals compete with each other with their fitness values. Here, the transformed unconstrained objective function due to Big-M and PFP penalty are considered as the fitness function. In the proposed GA, the well-known tournament selection process is employed as the selection operator. The primary objective of this process is to emphasize the above average solutions and eliminate the below average solutions from the population for the next generation under the well-known evolutionary principle "Survival of the fittest". This selection procedure is based on the following assumptions:

1. When both the chromosomes / individuals are feasible then the one with better fitness value is selected.
2. When one chromosome/individual is feasible and another is infeasible then the feasible one is selected.
3. When both the chromosomes/individuals are infeasible with unequal constraint violations, then the chromosome with less constraint violation is selected.
4. When both the chromosomes/individuals are infeasible with equal constraint violations, then any one chromosome/individual is selected.

After the selection process, new offspring will be created through crossover and mutation processes. In this work, we have used uniform crossover and uniform mutation in the genes corresponding to the integer variables, whole arithmetical crossover and boundary mutation for the last gene of the chromosome.

The computational steps of crossover are as follows:

Step-1: Find the integral value of the product of p_cross and p_size and store it in N.

Step-2: Select two chromosomes v_k and v_i randomly from the population.

Step-3: For first q genes, compute the components $\bar{x}_{kj}$ and $\bar{x}_{ij}$ $(j = 1, 2, ..., q)$ of two offspring by either $\bar{x}_{kj} = x_{kj} - g$ and $\bar{x}_{ij} = x_{ij} + g$ if $x_{kj} > x_{ij}$ or, $\bar{x}_{kj} = x_{kj} + g$ and $\bar{x}_{ij} = x_{ij} - g$, where g is a random integer number between 0 and $|x_{kj} - x_{ij}|$, $j = 1, 2, ..., q$ and for the last gene, compute the last components x'_{kj} and x'_{ij} of two offspring will be created by $x'_{kj} = cx_{kj} + (1-c)x_{ij}$ and $x'_{ij} = (1-c)x_{kj} + cx_{ij}$ where c is a random number between 0 and 1.

Step-4: Repeat step-2 and step-3 for $\frac{N}{2}$ times.

The computational steps of mutation are as follows:

Step-1: Find the integral value of the product of *p_mute* and *p_size* and store it in N.

Step-2: Select a chromosome v_i randomly from the population.

Step-3: Select a particular gene v_{ik} $(k = 1, 2, ..., q)$ of chromosome v_i for mutation and domain of v_{ik} is $[l_{ik}, u_{ik}]$.

Step-4: Create new gene v'_{ik} corresponding to the selected gene v_{ik} by mutation process as follows:

For $k = 1, 2, ..., q$

$$v'_{ik} = \begin{cases} v_{ik} + \Delta(u_{ik} - v_{ik}), & \text{if random digit is 0} \\ v_{ik} - \Delta(v_{ik} - l_{ik}), & \text{if random digit is 1} \end{cases}$$

$\Delta(y)$ returns a value in the range $[0, y]$, is a random integer between $[0, y]$.

Otherwise $v'_{ik} = \begin{cases} l_{ik} & \text{if a random digit is 0.} \\ u_{ik} & \text{if a random digit is 1.} \end{cases}$

Step-5: Repeat Step-2 to Step-4 for N times.

Sometimes, in any generation, there is a chance that the best chromosome may be lost when a new population is created by crossover and mutation operations. To remove this situation the worst individual/chromosome is replaced by the best individual/chromosome in the current generation. This process is called elitism. The different steps of this algorithm are described as follows:

7.1 Algorithm

Step 1: Initialize the parameters of genetic algorithm, bounds of variables and different parameters of the problem.

Step 2: $t = 0$ [t represents the number of current generation].

Step 3: Initialize the chromosome of the population$P(t)$[$P(t)$ represents the population at $t - th$ generation].

Step 4: Evaluate the fitness function of each chromosome of $P(t)$ considering any one of the objective function from (1-6) as fitness function.

Step 5: Find the best chromosome from the population $P(t)$.

Step 6: t is increased by unity.

Step 7: If the termination criterion is satisfied go to step-14, otherwise, go to next step.
Step 8: Select the population $P(t)$ from the population$P(t-1)$ of earlier generation by tournament selection process.
Step 9: Alter the population $P(t)$ by crossover, mutation and elitism process.
Step 10: Evaluate the fitness function value of each chromosome of $P(t)$.
Step 11: Find the best chromosome from $P(t)$.
Step 12: Compare the best chromosome of $P(t)$ and $P(t-1)$ and store better one.
Step 13: Go to step-6.
Step 14: Print the last found best chromosome (which is the solution of the optimization problem).
Step 15: End.

8 Numerical Example

To illustrate the proposed GAs (viz. PFP-GA and Big-M-GA) for solving earlier mentioned optimization problems with interval valued reliabilities of components, we have solved nine numerical examples. It is to be noted that for solving the said problem with fixed valued reliabilities of components, the reliability of each component is taken as interval with the same lower and upper bounds of interval. For each example, 20 independent runs have been performed by both the GAs, of which the following measurements have been collected to compare the performances of PFP-GA and Big-M-GA.

1. Best found system reliability
2. Average generations
3. Average CPU times

The proposed Genetic Algorithms are coded in C programming language and run in Linux environment. The computational work has been done on the PC which has Intel core-2 duo processor with 2 GHz. In this computation, different population size has been taken for different problems. However, the crossover and mutation rates are taken as 0.95 and 0.15 respectively.

Example 1: (related to Problem-1):

Maximize $[R_{SL}, R_{SR}] = \prod_{j=1}^{5} [\{1-(1-r_{jL})^{x_j}\}, \{1-(1-r_{jR})^{x_j}\}]$ subject to:

$$\sum_{j=1}^{5} p_j x_j^2 - P \leq 0,$$

$$\sum_{j=1}^{5} c_j [x_j + \exp(\tfrac{x_j}{4})] - C \leq 0,$$

$$\sum_{j=1}^{5} w_j x_j \exp(\tfrac{x_j}{4}) - W \leq 0,$$

The values of different parameters along with the interval valued reliabilities of Example-1 are given in Table 1.

Example 2: (related to Problem-2)
Maximize $[R_{SL}, R_{SR}] = \{1 - \langle 1 - (1 - [Q_{3L}, Q_{3R}](1 - [R_{1L}, R_{1R}][R_{2L}, R_{2R}]))[R_{4L}, R_{4R}]\rangle (1 - [R_{5L}, R_{5R}][R_{6L}, R_{6R}])\}(1 - [Q_{7L}, Q_{7R}][Q_{8L}, Q_{8R}][Q_{9L}, Q_{9R}])[R_{10L}, R_{10R}]$
subject to
$c_1 \exp(\frac{x_1}{2})x_2 + c_2 \exp(\frac{x_3}{2}) + c_3 x_4 + c_4[x_5 + \exp(\frac{x_5}{4})] + c_5 x_6^2 x_7 + c_6 x_8 + c_7 x_9^3 \exp(\frac{x_{10}}{2}) - 120 \leq 0,$
$w_1 x_1^2 x_2 + w_2 \exp(\frac{x_3 x_4}{2}) + w_3 x_5 \exp(\frac{x_6}{4}) + w_4 x_7 x_8^3 + w_5[x_9 + \exp(\frac{x_9}{2})] + w_6 x_2 \exp(\frac{x_{10}}{4}) - 130 \leq 0,$

Table 1 Parameters in Example 1

j	$[r_{jL}, r_{jR}]$	p_j	P	c_j	C	w_j	W
1	[0.76, 0.83]	1		7		7	
2	[0.82, 0.87]	2	110	7	175	8	200
3	[0.88, 0.93]	3		5		8	
4	[0.61, 0.67]	4		9		6	
5	[0.70, 0.80]	2		4		9	

The values of different parameters along with the interval valued reliabilities of Example-2 are given in Table 2.

Table 2 Parameters in Example 2

j	$[r_{jL}, r_{jR}]$	c_j	w_j	l_j	u_j
1	[0.80, 0.84]	8	16	1	4
2	[0.87, 0.90]	4	6	1	5
3	[0.89, 0.93]	2	7	1	6
4	[0.84, 0.86]	2	12	1	7
5	[0.88, 0.90]	1	7	1	5
6	[0.90, 0.95]	6	1	1	5
7	[0.80, 0.85]	2	9	1	3
8	[0.90, 0.95]	8	–	1	3
9	[0.80, 0.83]	–	–	1	4
10	[0.88, 0.92]	–	–	1	6

Example 3: (related to Problem 2)
Maximize
$[R_{SL}, R_{SR}] = \{1 - \langle 1 - (1 - [Q_{3L}, Q_{3R}](1 - [R_{1L}, R_{1R}][R_{2L}, R_{2R}]))[R_{4L}, R_{4R}]\rangle(1 - [R_{5L}, R_{5R}][R_{6L}, R_{6R}])\}(1 - [Q_{7L}, Q_{7R}][Q_{8L}, Q_{8R}][Q_{9L}, Q_{9R}])[R_{10L}, R_{10R}]$

subject to
$c_1 \exp(\frac{x_1}{2})x_2 + c_2 \exp(\frac{x_3}{2}) + c_3x_4 + c_4[x_5 + \exp(\frac{x_5}{4})] + c_5x_6^2x_7 + c_6x_8 + c_7x_9^3 \exp(\frac{x_{10}}{2}) - 120 \leq 0,$
$w_1x_1^2x_2 + w_2 \exp(\frac{x_3x_4}{2}) + w_3x_5 \exp(\frac{x_6}{4}) + w_4x_7x_8^3 + w_5[x_9 + \exp(\frac{x_9}{2})] + w_6x_2 \exp(\frac{x_{10}}{4}) - 130 \leq 0,$

The values of different parameters along with the interval valued reliabilities of Example-3 are given in Table 3.

Table 3 Parameters in Example -3

j	$[r_{jL}, r_{jR}]$	c_j	w_j	l_j	u_j
1	[0.83, 0.83]	8	16	1	4
2	[0.89, 0.89]	4	6	1	5
3	[0.92, 0.92]	2	7	1	6
4	[0.85, 0.85]	2	12	1	7
5	[0.89, 0.89]	1	7	1	5
6	[0.93, 0.93]	6	1	1	5
7	[0.83, 0.83]	2	9	1	3
8	[0.94, 0.94]	8	−	1	3
9	[0.82, 0.82]	−	−	1	4
10	[0.91, 0.91]	−	−	1	6

Example 4: (related to Problem-3)
Maximize $[R_{SL}, R_{SR}] = [R_{5L}, R_{5R}](1 - [Q_{1L}, Q_{1R}][Q_{3L}, Q_{3R}])(1 - [Q_{2L}, Q_{2R}][Q_{4L}, Q_{4R}]) + [Q_{5L}, Q_{5R}]\{1 - (1 - [R_{1L}, R_{1R}][R_{2L}, R_{2R}])(1 - [R_{3L}, R_{3R}][R_{4L}, R_{4R}])\}$ subject to:

$10 \exp(\frac{x_1}{2})x_2 + 20x_3 + 3x_4^2 + 8x_5 - 200 \leq 0,$
$10 \exp(\frac{x_1}{2}) + 4 \exp(x_2) + 2x_3^3 + 6[x_4^2 + \exp(\frac{x_4}{4})] + 7 \exp(\frac{x_5}{4}) - 310 \leq 0,$
$12[x_2^2 + \exp(x_2)] + 5x_3 \exp(\frac{x_3}{4}) + 3x_1x_4^2 + 2x_5^3 - 520 \leq 0,$
$(1, 1, 1, 1, 1) \leq (x_1, x_2, x_3, x_4, x_5) \leq (6, 3, 5, 6, 6),$
where
$R_1(x_1)$
$= \{[0.78, 0.82], [0.83, 0.88], [0.89, 0.91], [0.915, 0.935], [0.94, 0.96], [0.965, 0.985]\};$
$R_2(x_2) = 1 - (1 - [0.73, 0.77])^{x_2};$
$R_3(x_3) = \sum_{k=2}^{x_3+1} \binom{x_3 + 1}{k} ([0.87, 0.89])^k ([0.11, 0.13])^{x_3+1-k};$
$R_4(x_4) = 1 - (1 - [0.68, 0.72])^{x_4};$
$R_5(x_5) = 1 - (1 - [0.83, 0.86])^{x_5};$

Example 5: (related to Problem-3)
Maximize $[R_{SL}, R_{SR}] = [R_{5L}, R_{5R}](1 - [Q_{1L}, Q_{1R}][Q_{3L}, Q_{3R}])(1 - [Q_{2L}, Q_{2R}][Q_{4L}, Q_{4R}]) + [Q_{5L}, Q_{5R}]\{1 - (1 - [R_{1L}, R_{1R}][R_{2L}, R_{2R}])(1 - [R_{3L}, R_{3R}][R_{4L}, R_{4R}])\}$ subject to

$10\exp(\frac{x_1}{2})x_2 + 20x_3 + 3x_4^2 + 8x_5 - 200 \leq 0,$
$10\exp(\frac{x_1}{2}) + 4\exp(x_2) + 2x_3^3 + 6[x_4^2 + \exp(\frac{x_4}{4})] + 7\exp(\frac{x_5}{4}) - 310 \leq 0,$
$12[x_2^2 + \exp(x_2)] + 5x_3\exp(\frac{x_3}{4}) + 3x_1x_4^2 + 2x_5^3 - 520 \leq 0,$
$(1,1,1,1,1) \leq (x_1,x_2,x_3,x_4,x_5) \leq (6,3,5,6,6),$
where
$R_1(x_1)$
$= \{[0.8,0.8],[0.85,0.85],[0.9,0.9],[0.925,0.925],[0.95,0.95],[0.975,0.975]\};$
$R_2(x_2) = 1-(1-[0.75,0.75])^{x_2};$
$R_3(x_3) = \sum_{k=2}^{x_3+1}\binom{x_3+1}{k}([0.88,0.88])^k([0.12,0.12])^{x_3+1-k};$
$R_4(x_4) = 1-(1-[0.7,0.7])^{x_4};$
$R_5(x_5) = 1-(1-[0.85,0.85])^{x_5};$

The examples 1, 2, 3, 4 and 5 have been solved by two different methods PFP-GA and Big-M-GA and the results have been shown in Table 4.

Table 4 Numerical results for Example 1-5

Method	Exam -ple	Popul -ation size	x	Best found system reliability R_S	Average CPU seconds	Average Genera -tion
PFP -GA	1	50	$(3,2,2,3,3)$	[0.860808, 0.930985]	0.0001	12.10
	2	100	$(1,2,2,5,4,4,2,2,1,5)$	[0.999909, 0.999987]	0.0105	17.55
	3	100	$(1,2,2,5,4,4,2,2,1,5)$	[0.999975, 0.999975]	0.0100	17.55
	4	200	$(5,1,2,4,4)$	[0.991225, 0.999872]	0.0200	11.20
	5	100	$(3,2,4,4,2)$	[0.999382, 0.999382]	0.0100	12.40
Big-M -GA	1	50	$(3,2,2,3,3)$	[0.860808, 0.930985]	0.0001	12.80
	2	100	$(1,2,2,5,4,4,2,2,1,5)$	[0.999909, 0.999987]	0.0110	17.75
	3	100	$(1,2,2,5,4,4,2,2,1,5)$	[0.999975, 0.999975]	0.0100	17.75
	4	200	$(5,1,2,4,4)$	[0.991225, 0.999872]	0.0200	10.90
	5	100	$(3,2,4,4,2)$	[0.999382, 0.999382]	0.0100	12.55

Example 6: (related to the Problem-4)
Maximize $R_S(x,R) = R_1R_2 + Q_2R_3R_4 + Q_1R_2R_3R_4 + R_1Q_2Q_3R_4R_5$ $+Q_1R_2R_3Q_4R_5$ subject to:

$$C_1(x) = x_1x_2 + 2.2x_2x_3 + 1.5x_2x_4 + 2\exp\left(\frac{0.01}{1-R_5}\right) \leq 28,$$
$$C_2(x) = x_1 + 0.1x_2 + 2x_3 + x_4 + 5\exp\left(\frac{0.01}{1-R_5}\right) \leq 25,$$
$$C_3(x) = x_1^2 + (x_2-2)^3 + 1.5x_3 + x_4 + 0.6\exp\left(\frac{0.01}{1-R_5}\right) < 21,$$

where $1 \leq x_i \leq 6$, and are integers, $i = 1, 2, 3, 4$, $0.50 \leq R_5 \leq 0.99$,
and $R_i = R_i(x_i) = 1 - (1 - r_i)^{x_i}, i = 1, 2, 3, 4$, $Q_i = 1 - R_i, i = 1, ..., 5$
$r_1 = [0.69, 0.72]$,$r_2 = [0.83, 0.86]$,$r_3 = [0.73, 0.76]$,$r_4 = [0.79, 0.81]$

Example 7: (related to the Problem-4)
Maximize $R_S(x, R) = R_1R_2 + Q_2R_3R_4 + Q_1R_2R_3R_4 + R_1Q_2Q_3R_4R_5 + Q_1R_2R_3Q_4R_5$ subject to:

$$C_1(x) = x_1x_2 + 2.2x_2x_3 + 1.5x_2x_4 + 2\exp\left(\frac{0.01}{1-R_5}\right) \leq 28,$$

$$C_2(x) = x_1 + 0.1x_2 + 2x_3 + x_4 + 5\exp\left(\frac{0.01}{1-R_5}\right) \leq 25,$$

$$C_3(x) = x_1^2 + (x_2 - 2)^3 + 1.5x_3 + x_4 + 0.6\exp\left(\frac{0.01}{1-R_5}\right) < 21,$$

where $1 \leq x_i \leq 6$, and are integers, $i = 1, 2, 3, 4$, $0.50 \leq R_5 \leq 0.99$,
and $R_i = R_i(x_i) = 1 - (1 - r_i)^{x_i}, i = 1, 2, 3, 4$, $Q_i = 1 - R_i, i = 1, ..., 5$
$r_1 = [0.70, 0.70]$,$r_2 = [0.85, 0.85]$,$r_3 = [0.75, 0.75]$,$r_4 = [0.80, 0.80]$

The examples 6 and 7 have been solved by two different methods PFP-GA and Big-M-GA and the results have been shown in Table 5.

Table 5 Numerical results for Examples 6-7

Method	Example	Population size	(x, R)	Best found system reliability R_S	Average CPU seconds
PFP-GA	6	150	$(2, 3, 1, 2, 0.9900)$	$[0.958412, 0.997223]$	0.2705
	7	150	$(2, 1, 6, 5, 0.9396)$	$[0.999927, 0.999927]$	0.2655
Big-M-GA	6	150	$(2, 3, 1, 2, 0.9900)$	$[0.958412, 0.997223]$	0.2700
	7	150	$(2, 1, 6, 5, 0.9396)$	$[0.999927, 0.999927]$	0.2590

Example 8: (related to the Problem-5)
Minimize $C_w(x, R) = 0.3C_1(x_1) + 0.5C_2(x_2) + 0.2C_3(x_3)$ subject to: $R_S(x, R) \geq [0.999, 0.999]$, where $1 \leq x_i \leq 6$, and are integers, $i = 1, 2, 3, 4$, $0.50 \leq R_5 \leq 0.99$, and $R_S(x, R)$, $C_i(i = 1, 2, 3)$ are defined in Example 6.

Example 9: (related to the Problem-5)
Minimize $C_w(x, R) = 0.3C_1(x_1) + 0.5C_2(x_2) + 0.2C_3(x_3)$ subject to: $R_S(x, R) \geq [0.999, 0.999]$, where $1 \leq x_i \leq 6$, and are integers, $i = 1, 2, 3, 4$, $0.50 \leq R_5 \leq 0.99$, and $R_S(x, R)$, $C_i(i = 1, 2, 3)$ are defined in Example 7.

The examples 8 and 9 have been solved by two different methods PFP-GA and Big-M-GA and the results have been shown in Table 6.

Table 6 Numerical results for Example 8-9

Method	Exam -ple	Popu- lation size	(x, R)	Best found system cost C_w	Best found system reliability R_S	Average CPU seconds
PFP	8	150	$(6, 4, 2, 1, 0.8601)$	33.03866	$[0.997290, 0.999885]$	0.3675
-GA	9	150	$(2, 1, 4, 4, 0.5)$	17.97505	$[0.999081, 0.999081]$	0.3525
Big-M	8	150	$(6, 4, 2, 1, 0.8601)$	33.03866	$[0.997290, 0.999885]$	0.3010
-GA	9	150	$(2, 1, 4, 4, 0.5)$	17.97505	$[0.999081, 0.999081]$	0.2815

Table 7 Comparison of results of Ha and Kuo [1] and the proposed methods.

	Example	x	System Reliability R_s	Average CPU seconds
Ha and Kuo [1]	4(E2)	$(1, 1, 3, 4, 2, 1, 1, 3, 1, 4)$	0.999876	–
PFP-GA(this work)	3	$(1, 2, 2, 5, 4, 4, 2, 2, 1, 5)$	**0.999975**	**0.0100**
Big-M-GA(this work)	3	$(1, 2, 2, 5, 4, 4, 2, 2, 1, 5)$	**0.999975**	**0.0100**
Ha and Kuo [1]	4(E1)	$(1, 3, 4, 3, 3)$	0.999373	–
PFP-GA(this work)	5	$(3, 2, 4, 4, 2)$	**0.999382**	**0.0100**
Big-M-GA(this work)	5	$(3, 2, 4, 4, 2)$	**0.999382**	**0.0100**

Table 8 Comparison of results of Sun et al. [9] and the proposed methods.

	Example	(x, R)	System cost C_w	System Reliability R_s	Average CPU seconds
Sun et al.[9]	2	$(2, 1, 6, 5, 0.9396)$		0.99992653	9.84
PFP-GA(this work)	7	$(2, 1, 6, 5, 0.9396)$		**0.999927**	**0.2655**
Big-M-GA(this work)	7	$(2, 1, 6, 5, 0.9396)$		**0.999927**	**0.2590**
Sun et al.[9]	4	$(1, 1, 5, 4, .05)$	18.53505		15.97
PFP-GA(this work)	9	$(2, 1, 4, 4, 0.5)$	**17.97505**		**0.3525**
Big-M-GA(this work)	9	$(2, 1, 4, 4, 0.5)$	**17.97505**		**0.2815**

9 Sensitivity Analysis

To study the performance of our proposed GAs like PFP-GA and Big-M-GA based on two different types of penalty techniques, sensitivity analyses (for Example-1) have been carried out graphically on the centre of the interval valued system reliability with respect to GA parameters like, population size, crossover and mutation rates separately keeping the other parameters at their

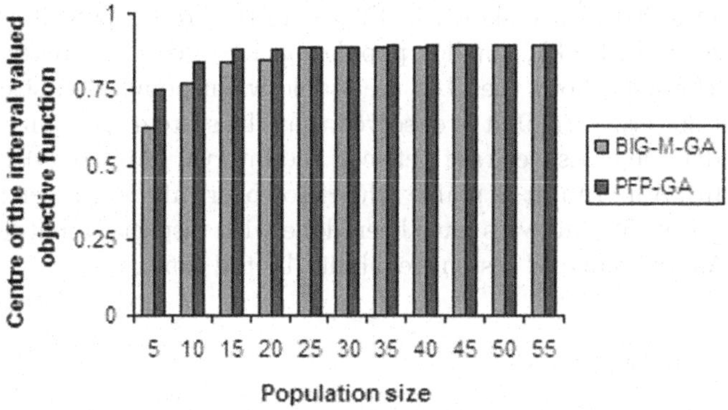

Fig. 8 Population size vs. centre of the objective function value

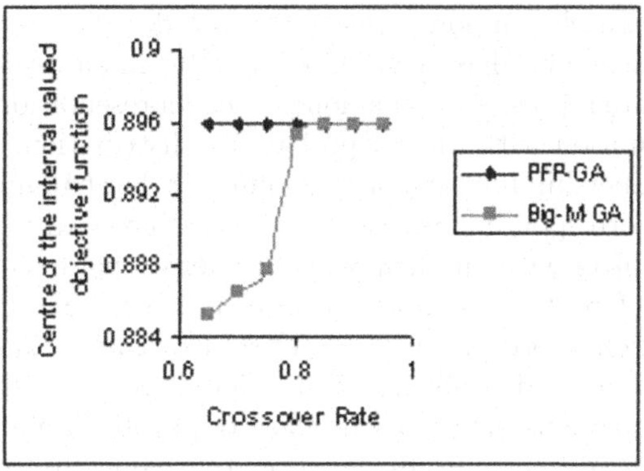

Fig. 9 Crossover rate vs.centre of the objective function value

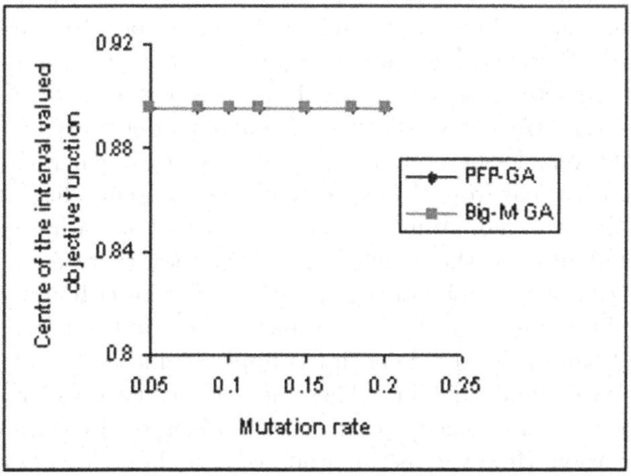

Fig. 10 Mutation rate vs.centre of the objective function value

original values. These are shown in Fig.8–Fig.10. From Fig.8, it is evident that in case of PFP-GA, smaller population size gives the better system reliability. However, both the GAs are stable when population size exceeds the number 30. From Fig.9, it is observed that the system reliability is stable if we consider the crossover rate between the interval (0.65, 0.95) in case of PFP-GA. In both GAs, it is stable when crossover rate is greater than 0.8. In Fig.10, sensitivity analyses have been done with respect to mutation rate. In both GAs, the value of system reliability be the same.

10 Conclusions

In this chapter, the problems of redundancy allocation problems of series system, hierarchical series-parallel system, complicated system and reliability network system with some resource constraints have been solved. In those systems, reliability of each component has been considered as imprecise number and this imprecise number has been represented by an interval number which is more appropriate representation among other representations like, random variable representation with known probability distribution, fuzzy set with known fuzzy membership function or fuzzy number. For handling the resource constraints, the corresponding problem has been converted to unconstrained optimization problem with the help of two different parameter free penalty techniques. Therefore, the transformed problem is of unconstrained interval valued optimization problem with integer and/or mixed integer variables. To solve the transformed problems, two different real coded GA based on different fitness functions have been developed for integer and mixed integer variables with interval valued fitness function, tournament selection, crossover (uniform crossover for integer variables and whole arithmetical crossover for floating point variables), mutation (uniform mutation for integer variables and boundary mutation for floating point variables) and elitism of size one. In the existing penalty function technique, tuning of penalty parameter is a formidable task. However, here tuning of parameters is not required as these are penalty parameter free techniques. From the performance of GAs, it is observed that the GAs with both fitness functions due to different penalty techniques take very lesser CPU times with very small generations to solve the problems. It is clear from the expression of the system reliability that the system reliability is a monotonically increasing function with respect to the individual reliabilities of the components. Therefore, there is one optimum setup irrespective of the choice of the upper bound or lower bound of the component reliabilities. As a result, the optimum setup obtained from the upper bound/lower bound will provide both the upper bound and the lower bound of the optimum system reliability. These approaches have wider applicabilities in solving the constrained optimization problems arisen in every sector of real life situation. However, as the proposed techniques are parameter free, these do not require the tuning of penalty parameter.

References

1. Ha, C., Kuo, W.: Reliability redundancy allocation: An improved realization for nonconvex nonlinear programming problems. European Journal of Operational Research 171, 124–138 (2006)
2. Coelho, L.S.: An efficient particle swarm approach for mixed-integer programming in reliability redundancy optimization applications. Reliability Engineering and System Safety 94, 830–837 (2009)
3. Tillman, F.A., Hwuang, C.L., Kuo, W.: Optimization technique for system reliability with redundancy: A Review. IEEE Trans. Reliability 26, 148–155 (1977)
4. Kuo, W., Prasad, V.R., Tillman, F.A., Hwuang, C.L.: Optimal Reliability Design Fundamentals and application. Cambridge University Press, Cambridge (2001)
5. Misra, K.B., Sharama, U.: An efficient algorithm to solve integer-programming problems arising in system reliability design. IEEE Trans. Reliability 40, 81–91 (1991)
6. Nakagawa, Y., Nakashima, K., Hattori, Y.: Optimal reliability allocation by branch-and-bounded technique. IEEE Trans. Reliability 27, 31–38 (1978)
7. Ohtagaki, H., Nakagawa, Y., Iwasaki, A., Narihisa, H.: Smart greedy procedure for solving a nonlinear knapsacclass of reliability optimization problems. Mathl. Comput. Modeling 22, 261–272 (1995)
8. Sun, X., Duan, L.: Optimal Condition and Branch and Bound Algorithm for Constrained Redundancy Optimization in Series System. Optimization and Engineering 3, 53–65 (2002)
9. Sun, X.L., Mckinnon, K.I.M., Li, D.: A convexification method for a class of global optimization problems with applications to reliability optimization. Journal of Global Optimization 21, 185–199 (2001)
10. Gen, M., Yun, Y.: Soft computing approach for reliability optimization. Reliability Engineering and System Safety 91, 1008–1026 (2006)
11. Chern, M.S.: On the computational complexity of reliability redundancy allocation in a series system. Operations Research Letter 11, 309–315 (1992)
12. Martorell, S., Sanchez, A., Carlos, S., Serradell, V.: Alternatives and challenges in optimizing industrial safety using genetic algorithms. Reliability Engineering and System Safety 86, 25–38 (2004)
13. Zhao, J., Liu, Z., Dao, M.: Reliability optimization using multiobjective ant colony system approaches. Reliability Engineering and system safety 92, 109–120 (2007)
14. Zio, E.: Reliability engineering: Old problems and new challanges. Reliability Engineering and System Safety 94, 125–141 (2009)
15. Coolen, F.P.A., Newby, M.J.: Bayesian reliability analysis with imprecise prior probabilities. Reliability Engineering and System Safety 43, 75–85 (1994)
16. Utkin, L.V., Gurov, S.V.: Imprecise reliability of general structures. Knowledge and Information Systems 1(4), 459–480 (1999)
17. Utkin, L.V., Gurov, S.V.: New reliability models based on imprecise probabilities. In: Hsu, C. (ed.) Advanced Signal Processing Technology, pp. 110–139. World Scientific, Singapore (2001)
18. Gupta, R.K., Bhunia, A.K., Roy, D.: A GA based penalty function technique for solving constrained redundancy allocation problem of series system with interval valued reliability of components. Journal of Computational and Applied Mathematics 232, 275–284 (2009)

19. Bhunia, A.K., Sahoo, L., Roy, D.: Reliability stochastic optimization for a series system with interval component reliability via genetic algorithm. Applied Mathematics and Computation 216, 929–939 (2010)
20. Mahato, S.K., Bhunia, A.K.: Interval-Arithmetic-Oriented Interval Computing Technique for Global Optimization. In: Applied Mathematics Research eXpress 2006, pp. 1–19 (2006)
21. Ishibuchi, H., Tanaka, H.: Multiobjective programming in optimization of the interval objective function. European Journal of Operational Research 48, 219–225 (1990)
22. Chanas, S., Kuchta, D.: Multiobjective programming in the optimization of interval objective functions-A generalized approach. European journal of Operational Research 94, 594–598 (1996)
23. Goldberg, D.E.: Genetic Algorithms: Search, Optimization and Machine Learning. Addison-Wesley, Reading (1989)
24. Gen, M., Cheng, R.: Genetic algorithms and engineering optimization. John Wiley and Sons Inc., Chichester (2000)
25. Michalawich, Z.: Genetic Algorithms + Data structure = Evaluation Programs. Springer, Berlin (1996)
26. Sakawa, M.: Genetic Algorithms and fuzzy multiobjective optimization. Kluwer Academic Publishers, Dordrecht (2002)
27. Levitin, G.: Genetic algorithms in reliability engineering. Reliability Engineering and System Safety 91, 9751–9976 (2006)
28. Villanueva, J.F., Sanchez, A.I., Carlos, S., Martorell, S.: Genetic algorithm-based optimization of testing and maintenance under uncertain unavailability and cost estimation: A survey of strategies for harmonizing evoluation and accuracy. Reliability Engineering and System Safety 93, 1830–1841 (2008)
29. Ye, Z., Li, Z., Xie, M.: Some improvements on adaptive genetic algorithms for reliability-related applications. Reliability Engineering and System Safety 95, 120–126 (2010)
30. Deb, K.: An efficient constraint handling method for genetic algorithms. Computer Methods in Applied Mechanics and Engineering 186, 311–338 (2000)
31. Aggarwal, K.K., Gupta, J.S.: Penalty function approach in heuristic algorithms for constrained. IEEE Transactions on Reliability 54(3), 549–558 (2005)
32. Miettinen, K., Makela, M.M., Toivanen, J.: Numerical comparison of some Penalty-Based Constraint Handling Techniques in Genetic Algorithms. Journal of Global Optimization 2, 427–446 (2003)
33. Hansen, E., Walster, G.W.: Global optimization using interval analysis. Marcel Dekker Inc., New York (2004)
34. Karmakar, S., Mahato, S., Bhunia, A.K.: Interval oriented multi-section techniques for global optimization. Journal of Computation and Applied Mathematics 224, 476–491 (2009)

PSO in Building Fuzzy Systems

Nadia Nedjah[1], Sergio Oliveira Costa Jr.[1], Luiza de Macedo Mourelle[2], Leandro dos Santos Coelho[3], and Viviana Cocco Mariani[3]

[1] Department of Electronics Engineering and Telecommunications
[2] Department of Systems Engineering and Computation
Faculty of Engineering, State University of Rio de Janeiro
{nadia,serol,ldmm}@eng.uerj.br
[3] Department of Electrical Engineering, PPGEE Federal University of Parana, UFPR Polytechnic Center, Curitiba, Parana, Brazil

Summary. In this chapter, we take advantage of particle swarm optimization to build fuzzy systems automatically for different kinds of problems by simply providing the objective function and the problem variables. Particle swarm optimization (PSO) is a technique used in complex problems, including multi-objective problems. Fuzzy systems are currently used in many kinds of applications, such as control, for their effectiveness and efficiency. However, these characteristics depend primarily on the model yield by human experts, which may or may not be optimized for the problem. To avoid dealing with inconsistent during the fuzzy systems generation, we used some known techniques, such as the WM method, to help evolving meaningful rules and clustering concepts to generate membership functions. Tests using three three-dimensional functions have been carried out and show that the evolutionary process is promising.

1 Introduction

Fuzzy systems [5] form an important tool to model complex problems based on imprecise informations and/or in situations where a precise result is not of interest and an approximation is sufficient [16]. The performance of a fuzzy system depends on the expert's interpretation, which leads to in the generation of the rule base and membership functions of the system. To minimize this dependency, some methods are being used in the attempt to automatically generate the components required in a fuzzy system. For the membership functions, clustering-based algorithms, such as Fuzzy C-means and its generalizations as *Pre-shaped C-means* [2], are usually used. Other approaches also exist [10]. The major difficulty in the development of fuzzy systems consists of the definition of membership functions and rules that provide the desired behavior of these systems.

Swarm Intelligence is an area of artificial intelligence based on collective and decentralized behavior of individuals that interact with each other, as

N. Nedjah et al. (Eds.): Innovative Computing Methods, SCI 357, pp. 37–52.
springerlink.com

well as with the environment [1]. PSO is a stochastic evolutionary algorithm, based on swarm intelligence, that searches for the solution of optimization problems in a specific search space and is able to predict the social behavior of individuals according to defined objectives [6].

Methods based on examples, such as the Wang-Mendel or WM method [15], are usually used for automatic rule generation. Also, there are many research works that exploit evolutionary algorithms (EA), both to optimize the rule base and the membership functions. In [3], genetic algorithms (GA) are used to generate the rule base, with candidate rules pre-selection. In [9], the authors use EA to generate fuzzy systems that are more compact and more interpretable by humans. In [13], the authors use clustering techniques and GA to define good sets of rules for classification problems. In [4], the authors use evolutionary technique and GA to generate fuzzy systems from some given knowledge bases.

In this paper, we developed an algorithm based on PSO to generate fuzzy systems for any kind of problem, provided an objective function that may be continuous or discrete. Using simple informations, such as variable names, the corresponding domains and the objective function, this algorithm can yield an appropriate fuzzy system. Some tests were performed with a known control surface to validate the effectiveness of the tool.

The rest of this paper is organized in five sections. Firstly, in Section 2, we explain briefly the principle behind PSO. Then, in Section 3, we describe the WM method of rules generation. After that, in Section 4, we give details about the proposed method for the automatic modeling of fuzzy systems using PSO. For this purpose, we first define the structure of a particle and the coordinates used to position it within the search space as well as the fitness function of the represented system. Then, in Section 5, we present the obtained results to model a commonly used control surface. Last but not least, in Section 6, we conclude the reported work and give some directions for future research.

2 Particle Swarm Optimization

During a particle swarm optimization process, each particle is mapped into a position in the search space, which n-dimensional. The particle position is updated in each iteration. For the position update particle i, the velocity related to each of the search space directions is used. The velocity is the element that promotes the movement of the particles and is calculated as in (1) [11, 8, 6].

$$v_i(t+1) = wv_i(t) + c_1r_1(\hat{x}_i(t) - x_i(t)) + c_2r_2(\bar{x}(t) - x_i(t)), \qquad (1)$$

where w is called inertia coefficient, r_1 and r_2 are random numbers chosen in the interval [0,1], c_1 and c_2 are positives constants called as social and cognitive coefficients, $\hat{x}_i(t)$ identifies the best position achieved by the particle in the past and $\bar{x}(t)$ is the best position, among all the particles, achieved in the past. The position of the particle is updated as described in (2).

$$x_i(t+1) = x_i(t) + v_i(t+1). \qquad (2)$$

The velocity guides the optimization process [6], reflecting both the particle experience and the information exchange between particles. The experimental knowledge of a particle refers to the cognitive component, which is proportional to the distance between the particle and its best position, found so far. The information exchange between particles refers to the social component of the velocity equation (1).

To avoid that a particle leaves the search space, it is necessary to use a parameter that bounds the velocity [6]. This is known as the maximum velocity v_{max}, it allows a higher granularity on the search control. Therefore, before executing the update defined in (2), the velocity is analyzed with respect to the criterion defined in (3).

$$v_i(t+1) = \begin{cases} v_i(t+1) \text{ if } v_i(t+1) < v_{max} \\ v_{max} \qquad \text{ if } v_i(t+1) \geq v_{max} \end{cases} \qquad (3)$$

In (1), one can observe three terms that interfere in the velocity computation [6], which are:

- The *previous velocity*, $wv_i(t)$, is used to prevent particle i to suffer a drastic change in direction. This component also is called of inertia component.
- The *cognitive component*, $c_1r_1(\hat{x}_i(t) - x_i(t))$, quantifies the performance of particle i with respect to previous performances. This component was defined by Kennedy and Eberhart as the "nostalgia" of the particle [8].
- The *social component*, $c_2r_2(\bar{x}(t) - x_i(t))$, quantifies the performance of particle i with respect to the performance of the set of included particles. The effect of this term is to attract the particle to the best position found by the particles set.

The value assigned to each parameter of PSO algorithm is essential in the search process evolution. Below are related some set of values considered *good*.

- The *inertia coefficient*, w, controls the relation between exploration and exploitation [14]. The values near 1 are considered good, but values bigger than 1 are not so good so are very small values [6]. Values bigger than 1 tend to leave the particles with a very high acceleration, promoting a high divergence, while very small values can make the search too slow.
- The *cognitive coefficient*, c_1, and the *social coefficient*, c_2, yield a better performance when these are balanced, i.e., $c_1 \cong c_2$ [6].
- The factors r_1 and r_2 define the stochastic nature of the cognitive and social contributions. Random values are selected in the range [0,1] [6] to each factors.

- The *maximum velocity* is defined for each of the dimensions of search space and can be formulated as a domain percentage [6], $v_{max} = \delta(x_{max} - x_{min})$, where x_{max} and x_{min} are the maximum and minimum domain values respectively and δ is a value in the range [0, 1].
- The *number of particles* define the possibility to cover a range of the search space in every iteration of the algorithm. A high number of particles allows a better coverage of the search space but requires a considerable computational power. Empirical studies show that PSO achieves optimal solutions using ten to thirty particles [6].

Let*pBest* be the best position found by a particle and *gBest* be the best position among those found by all the particles. Algorithm 1 is describes the PSO optimization process. A given maximal iteration number and the predefined fitness values can be used as a stop criterion.

Algorithm 1. Particle swarm-based optimization algorithm (PSO)

```
for i := 1 until total_particulas do
    Initialize particle i information;
    Initialize random position of particle i;
    Initialize random velocity of particle i;
end for
repeat
    for i := 1 until total_particulas do
        Calculate fitness of particle i;
        if (fitness better than pBest_i) then
            Update pBest_i with the new position;
        end if
        if (fitness better than gBest) then
            Update gBest with the new position;
        end if
        Update velocity of particle i;
        Update position of particle i;
    end for
until (stopcriterion = true)
```

The main characteristic of the PSO algorithm is the social interaction [6], which makes the individuals able to learn with the group and use the acquired knowledge. This allows the particles follow the path that leads to more success, as it happens in nature.

A topology is the way that can be used to understand the interaction between a particle and the group in which it is inserted. In practical terms, a topology in PSO is defied by the manner the position update in performed. Therefore, it is defined by the way the velocity is computed because it through the velocity the knowledge about the best path is transmitted between the particles of the swarm [6] [11].

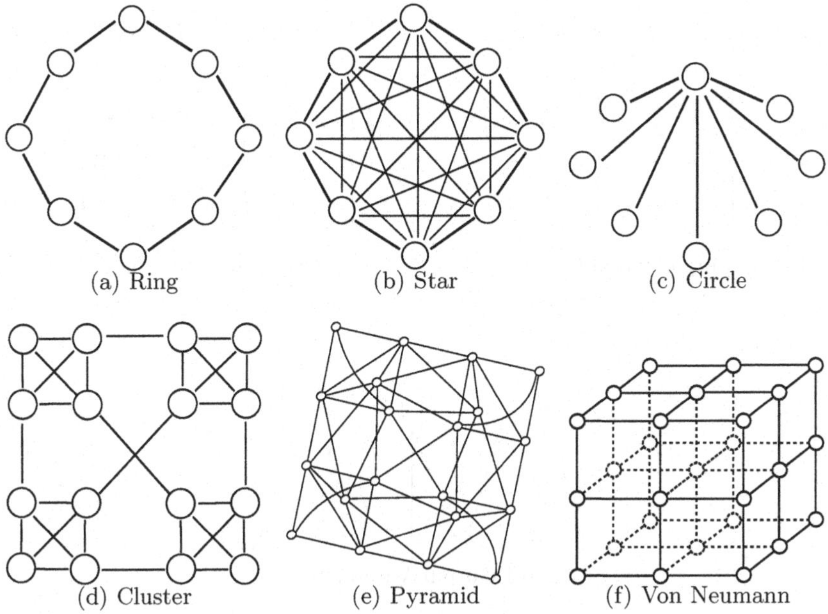

Fig. 1 Topologies for PSO

Figure 1(a) shows the ring topology, also known as *Local Best PSO*. In this topology, each particle is connected to two others, which are in its immediate neighborhood. Figure 1(b) shows the star topology, also known as *Global Best PSO*. In this topology, each particle is connected to all those of the swarm, i.e., a particle neighborhood is formed by all existing particles.

Besides these two types of topologies, Figure 1(c) shows all particles connected to central particle that controls the flow of information. In Figure 1(d), the particles are clustered and a each particle is considered neighbor of all those of its corresponding cluster, yet some of the cluster particle allow for a connection with the other in the neighboring clusters. In Figure 1(e), shows the pyramid topology which connects particles in triangles. Lastly, Figure 1(f) shows the Von Neumann structure, wherein the particles are connected via a grid structure [6].

3 Rule Generation Methods

The rule generation method referred to as Wang-Mendel (WM) [15] uses an input-output data set for the problem at hand, to generate a rule set of fuzzy systems. The input-output data set is usually provided as $(x^p; y^p)$, $p = 1, 2, \ldots, N$, wherein $x^p \in R^n$ and $y^p \in R$. This method extracts the rules that best describe how the output variable $y \in R$ is influenced by the n input variables $x = (x_1, ..., x_n) \in R^n$, based on the provided examples.

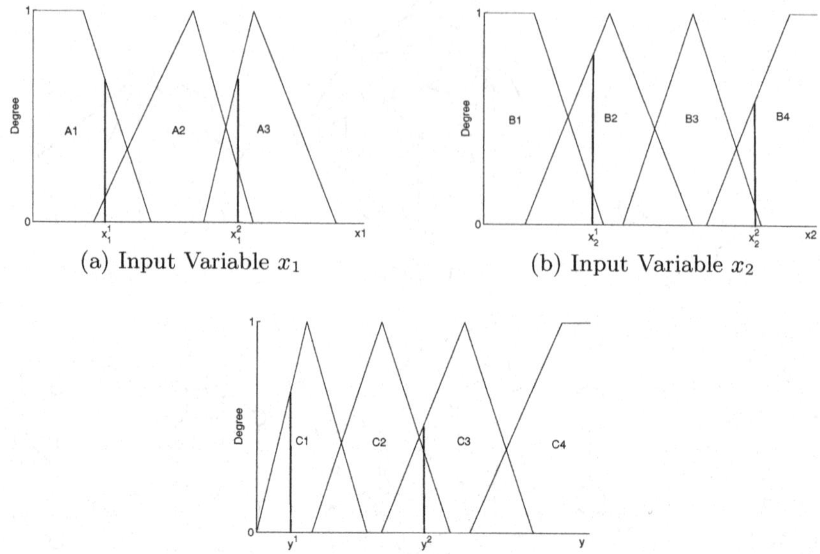

(a) Input Variable x_1 (b) Input Variable x_2

(c) Output Variable y

Fig. 2 Example of input-output data set for rules generation

For instance, assuming two data sets to a system with two input variables x_1 and x_2 and an output y, that are (x_1^1, x_2^1, y^1) and (x_1^2, x_2^2, y^2), and the membership functions showed in the graphics of the Fig. 2. To obtain the rules represented by these two sets, first we must get the degree of confidence using the membership functions, for each data set. In this case, we have:

- x_1^1: degree 0.67 in $A1$ and 0.11 in $A2$;
- x_2^1: degree 0.16 in $B1$ and 0.80 in $B2$;
- y^1: degree 0.66 in $C1$;
- x_1^2: degree 0.25 in $A2$ and 0.68 in $A3$;
- x_2^2: degree 0.10 in $B3$ and 0.58 in $B4$;
- y^2: degree 0.39 in $C2$ and 0.51 in $C3$.

In order to generate the rules, we always keep the membership functions in which the variable has the highest degree, and so we discard the functions that have lower degree. So, for the data sets defined in Fig. 2, we have (4). So the first rule would be "If x_1 is $A1$ and x_2 is $B2$ then y is $C1$" and the second, "If x_1 is $A3$ and x_2 is $B4$ then y is $C3$".

$$\begin{array}{l}(x_1^1, x_2^1, y^1) = [x_1^1(0.67 \text{ in } A1), x_2^1(0.80 \text{ in } B2), y^1(0.66 \text{ in } C1)] \\ x_1^2, x_2^2, y^2) = [x_1^2(0.68 \text{ in } A3), x_2^2(0.58 \text{ in } B4), y^2(0.51 \text{ in } C3)]\end{array} \quad (4)$$

Note that each data set generates one single rule. Considering a real system, it is very possible that these rules can be conflicting rules. To overcome this problem, one can associate degrees of confidence to each generated rule, using the degree of relevance of each rule term. Equation 5 shows how this degree can be computed:

$$C(Rule) = \mu(x_1) \times \mu(x_2) \times \mu(y), \tag{5}$$

wherein C is degree of *Rule* and $\mu(x1)$, $\mu(x2)$ and $\mu(y)$ are the degree of relevance of each rule term. In the case of the first rule, the associated confidence degree would be $C(Rule_1) = 0.67 \times 0.80 \times 0.66 = 0.35376$.

There are also methods based on genetic algorithms. In [3], the authors use the WM method to generate the initial rule set and then apply their own genetic algorithm on some classification rate of the rules. This method is only used for classification problems.

4 Proposed Automatic Generation

In this work, PSO is used to evolve the fuzzy systems parameters of the *Mamdani* type [5], both for rules and membership functions. The search algorithm is based on these two elements and always tries to improve the solution at hand. However, the functions are not modified in every iteration, unlike the rules, whose modification obeys to pre-determined update rate, that is defined at the beginning of the evolutionary process. The purpose is to maintain the functions stable for some time, giving more time for the algorithm to search for more appropriate rules for those functions. At the end of each execution, when the algorithm reaches the stop criterion, it returns the best solution found.

There are four important aspect that define the performance of the PSO search. These are the particle representation, the position coordinates of a given particle in the search space, the fitness function that allows us to determine how appropriate is the fuzzy system associated with a given particle and how to update the system represented by the particle at hand.

4.1 Representation

Each particle is associated with a fuzzy system and a position in the search space, that is represented by an n-position vector, where n depends on the number of used variables. In this work, the fuzzy system is defined by an hierarchical structure as described in Fig. 3.

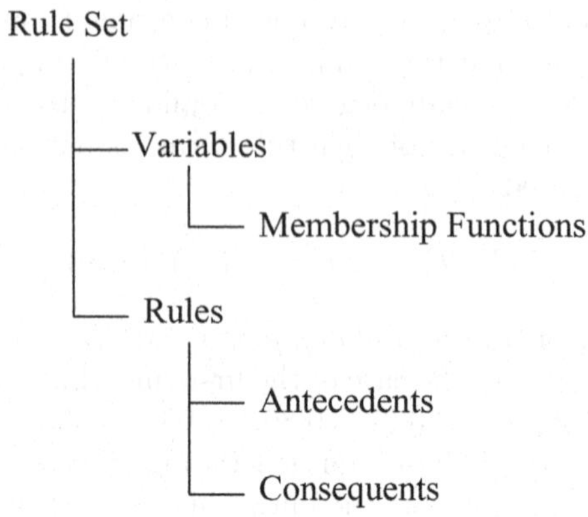

Fig. 3 Fuzzy representation structure

4.2 Particle Position and Movement

The position vector has one entry for the rules number, another one for the completeness factor, and m positions for the number of the functions, where m is the number of the system variables. Thus, the position dimension is dependent on the number of variables of the problem. The completeness factor is a criterion that measures the *discontinuity* between functions in the variables domain (see Section 4.3). The mutation operator determines how the update of each one these items is performed. This update promotes the movement of particles on the search space. In this work, we used three kinds of mutations:

- If the velocity relating to the number of rules is positive, then we increase the rules number. Otherwise, we decrease it.
- Changing the number of functions for each variable of the fuzzy system follows the same criterion, given above.
- The change of the completeness is performed increasing the width of a function. Thus, the tendency is to reduce the empty space in the domain, if any. Similarly, reducing the width of the function, we alter the distinctness between the available ones. The more positive the velocity is, the bigger the increase in domain of one of the functions. If the more negative the velocity is, the smaller the decrease in the domain.

4.3 Fitness Function

Inspired by the work reported in [9], the fitness of each particle is defined as in (6):

$$\begin{aligned}F = &-100 \times \omega_1 \times \log(MSE)\\ &+50 \times \omega_2(1 - C_r)\\ &+50 \times \omega_3(1 - C_f)\\ &+50 \times \omega_4(1 - P_D)\\ &+50 \times \omega_5(1 - P_C),\end{aligned} \tag{6}$$

wherein MSE is the mean-square error of the difference between the returned value by the objective function (y_h) and the returned value by the fuzzy model (y_h^F), as (7), wherein N_D is the number of data.

$$MSE = \frac{1}{N_D}\sum_{h=1}^{N}(y_h - y_h^F)^2, \tag{7}$$

The C_r term represents the relation between the amount of rules presents on the model and the total of possible rules, and C_f represents the relation between the amount of functions presents on the system and the total of possible functions, as in (8):

$$\begin{aligned}C_r &= \tfrac{N_R}{N_R^{max}}\\ C_f &= \tfrac{N_F}{N_F^{max}}\end{aligned} \tag{8}$$

Term P_D is a criterion that measures the *distinctness* between the membership functions of the variables, defined in (9):

$$P_D = \frac{1}{N_V}\sum_{i=1}^{N_V}\left(\frac{1}{N_S^i}\sum_{h=1}^{N_S^i}\frac{\lambda_{ih}}{|\chi_i|}\right), \tag{9}$$

wherein N_V is the number of variables, N_S^i is the total possible interval of overlap between functions of the i-th variable, λ_{ih} is the width of the h-th overlap and χ_i is the width of the variable domain. In order to the determine λ_{ih}, it is necessary to define the level ξ_D, drawing a horizontal line, crossing all the functions, as showed in the Fig. 4(a). Term P_C is a criterion that represents the *completeness* of the membership functions, in relation to the domain, and is defined as in (10):

$$P_C = \frac{1}{N_V}\sum_{i=1}^{N_V}\left(\frac{1}{N_D^i}\sum_{h=1}^{N_D^i}\frac{\gamma_{ih}}{|\chi_i|}\right), \tag{10}$$

wherein N_D^i is the total possible number of discontinuity between functions of the i-th variable, γ_{ih} is the width of h-th discontinuity and χ_i is the width of the variable domain.

In order to determine γ_{ih}, is necessary to define the level ξ_C, in a similar way to ξ_D, as shown Fig. 4(b).

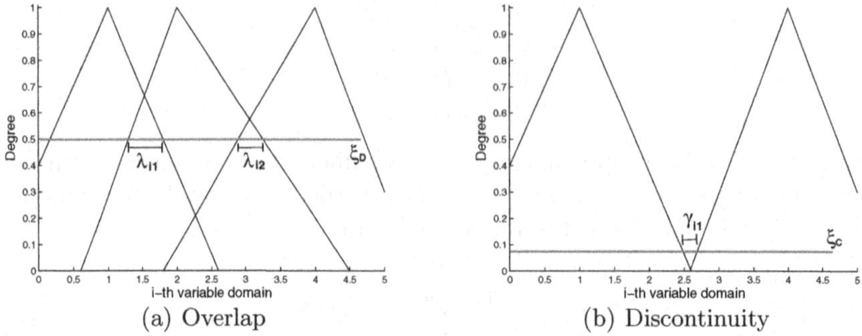

Fig. 4 Overlap and discontinuity illustrations

The coefficients ω_1, ω_2, ω_3, ω_4 and ω_5, control the contribution of each term of (6) in the evaluation of the fuzzy system associated with the particle.

This evaluation function covers the many required criteria of a fuzzy system. These are the *precision*, by the error quantification; the *compactness*, by the relation between the number of the rules and functions of the model and the total possible number; and the *interpretability*, by measuring of the distinctness and completeness, providing a complete model evaluation.

5 Results and Tests

The WM method [15], introduced in Section 3, was used to initialize rules of the fuzzy systems of each particle. Besides, the concept of clustering was used in the membership functions generation, to decrease the possibility of yielding functions that are incompatible with the fuzzy system.

Experiments were performed to prove the effectiveness of the technique. Table 1 show the used parameters by the PSO-based algorithm. Fig. 5 depicts the graphics of the original function $z = seno(xy)$. The best result obtained

Table 1 Values of the algorithm parameters

Parameter	Value
Social coefficient	1.5
Cognitive coefficient	1.5
Inertia coefficient	1
Number of particles	20
Completeness	0.25
Number of iteration	5000
Total of rules WM	6
Kind of function	Gaussian
$\omega_1 - \omega_5$	1

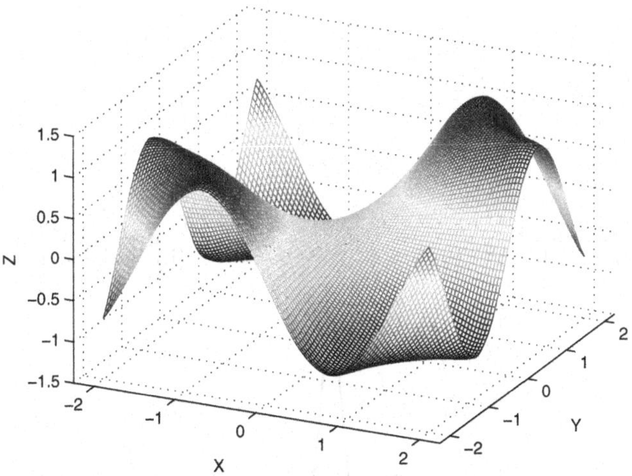

Fig. 5 Original function $z = seno(xy)$

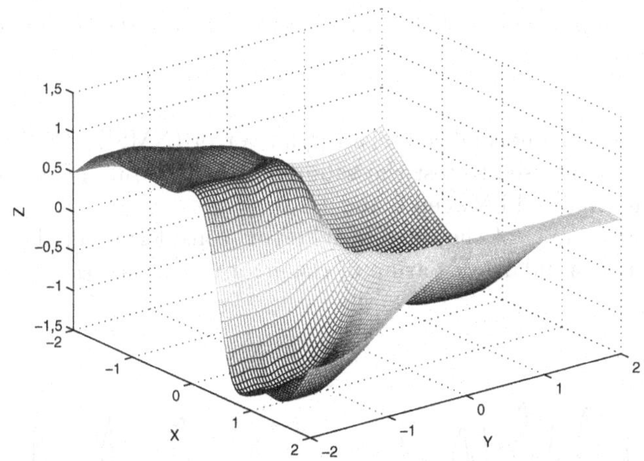

Fig. 6 Best result obtained for function $z = seno(xy)$

for the approximation of this function using the fuzzy system generated by the PSO-based algorithm is shown in Fig. 6.

Figure 7 shows the configuration input and output variables of the one of the systems yield by the PSO algorithm. The numbering of the membership function does not follow the order in which these appear in the domain. This due to the Java system used to program the process.

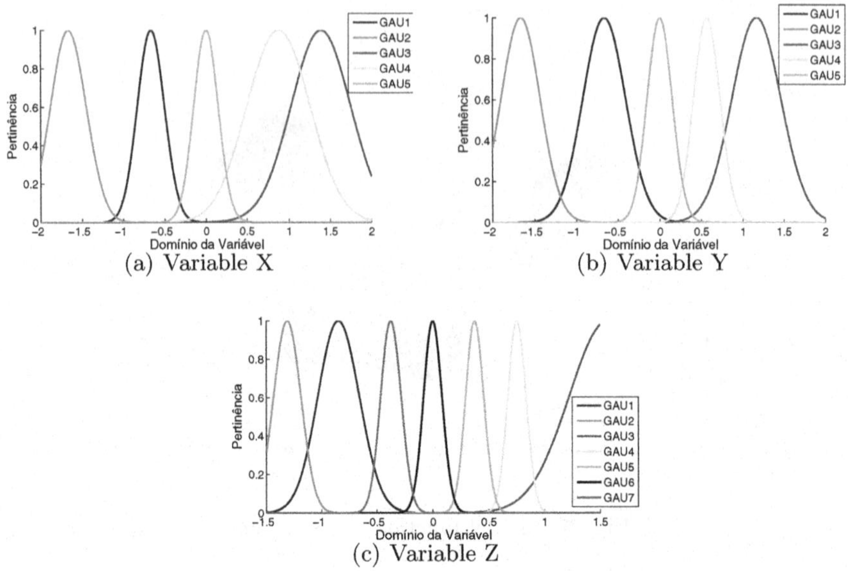

(a) Variable X (b) Variable Y

(c) Variable Z

Fig. 7 Membership functions of the fuzzy variables used in the generated system for function $seno(xy)$

Figure 8 shows the normalized mean square error (NMSE), that was computed for each point used to test the validity of the generated fuzzy systems. The error is normalized between 0 and 1.

The general average of the introduced error, considering all the points is $\nabla = 0.1359$. The standard deviation of the error is $\delta = 0.0537$.

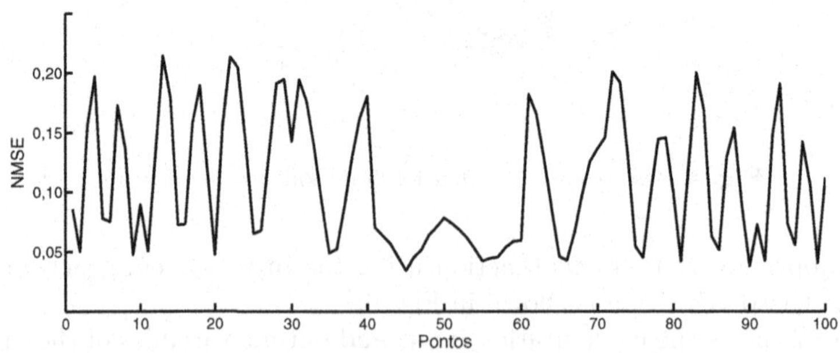

Fig. 8 Error averages for the fuzzy system for function $seno(xy)$

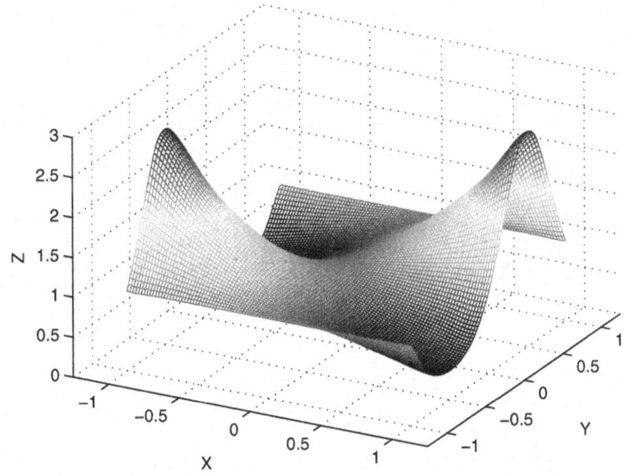

Fig. 9 Original function $z = e^{x\,sin(\pi y)}$

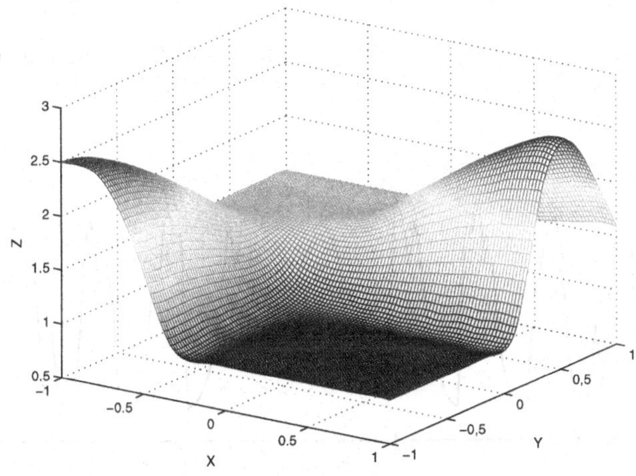

Fig. 10 Best result obtained for function $z = e^{x\,sin(\pi y)}$

Fig. 9 depicts the graphics of the original function $z = e^{x sin(\pi y)}$. The best result obtained for the approximation of this function using the fuzzy system generated by the PSO-based algorithm are shown in Fig. 10.

Figure 11 shows the configuration of the input and output variables of the resulting fuzzy system.

Figure 12 shows the average of the error introduced. In this case, the general error average is $\nabla = 0.2514$, while the standard deviation is $\delta = 0.1364$.

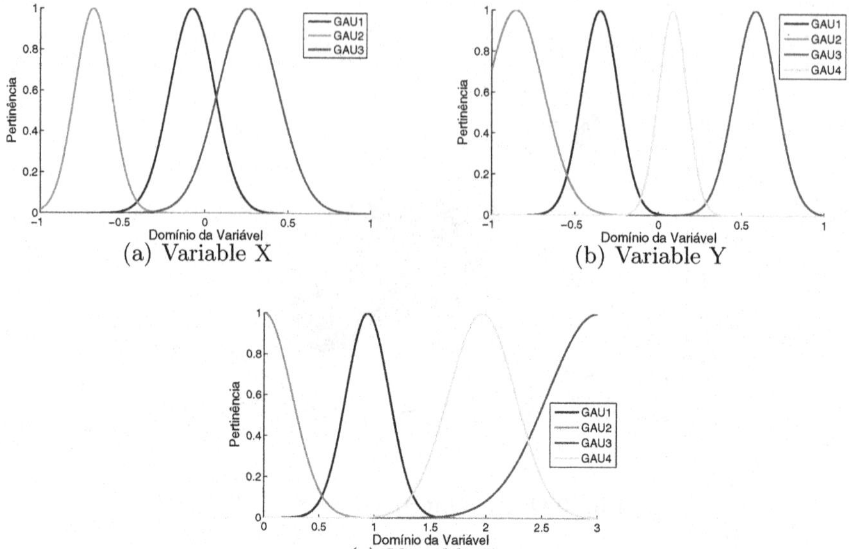

(a) Variable X (b) Variable Y

(c) Variable Z

Fig. 11 Membership functions of the fuzzy variables of the system generated for the function $z = e^{x sin(\pi y)}$

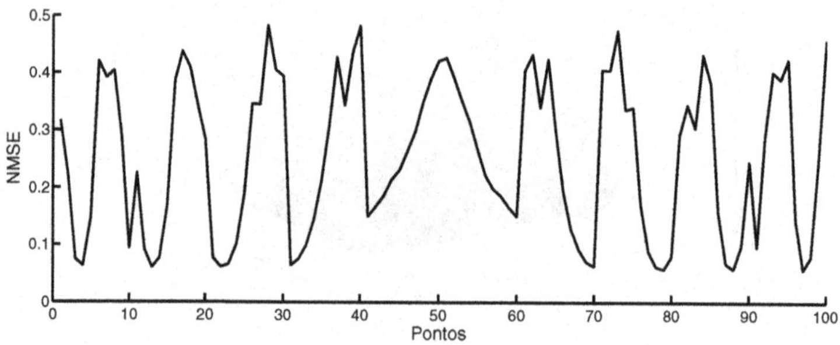

Fig. 12 Error averages for the fuzzy system for the function $z = e^{x sin(\pi y)}$

Table 2 shows a comparison between the results obtained by automatic generation of fuzzy systems using genetic algorithms [12] and using the proposed method using PSO. As before, the error is computed using the Normalized Mean Square Error (NMSE).

Table 2 Comparison of the performance of GA-based *vs.* PSO-based fuzzy system automatic generation for functions $seno(xy)$ and $e^{x sin(\pi y)}$

Function	Algorithm	Fitness	Deviation of fitness	Error	Deviation of error
$seno(xy)$	GA	54	8.0	0.0288	0.0007
	PSO	275.5	3.8	0.1059	0.0537
$e^{x sin(\pi y)}$	GA	197	35.0	0.006	0.0032
	PSO	280.5	16.8	0.176	0.1656

6 Conclusion

In this paper, we illustrated the use of PSO to automatically generate the fuzzy rules, fuzzy variables together with the corresponding membership functions of fuzzy systems. We described the particle representation, its movement in the search space and we provided a fitness function that allows us to assess the appropriateness of the evolved fuzzy system. This experience showed that the performance of the evolutionary process is very much dependent on the choice of the parameters, such as the number of membership functions per variable as well as on the number of rules allowed in the system. More tests are being carried out in order to synthesize discrete functions into fuzzy systems.

Acknowledgments

The authors are grateful to FAPERJ (*Fundação de Amparo à Pesquisa do Estado do Rio de janeiro*, `http:// www.faperj.br`), CNPq (*Conselho Nacional de Desenvolvimento Científico e Tecnológico*, `http://www.cnpq.br`) and CAPES (*Coordenação de Aperfeiçoamento de Pessoal de Nível Superior* `http://www.capes.gov.br`)for their continuous financial support.

References

1. Beni, G., Wang, J.: Robots and Biological Systems: Towards a New Bionics? Toscana, Italy. NATO ASI Series (1989)
2. Chen, L., Chen, C.L.P.: Pre-shaped fuzzy c-means algorithm (pfcm) for transparent membership function generation. In: Proc. of IEEE International Conference on Systems, Man and Cybernetics, pp. 789–794 (October 2007)
3. Cintra, M.E., Camargo, H.A.: Fuzzy rules generation using genetic algorithms with self-adaptive selection. In: Proc. of IEEE International Conference on Information Reuse and Integration, pp. 261–266 (August 2007)
4. Cordón, O., Herrera, F.: A hybrid genetic algorithm-evolution strategy process for learning fuzzy logic controller knowledge bases. In: Genetic Algorithms and Soft Computing, pp. 251–278. Physica-Verlag, Heidelberg (1996)

5. Cox, E.: The Fuzzy Systems Handbook: A Practitioner's Guide to Building, Using, and Maintaining Fuzzy Systems. Academic Press Limited, Oval Road (1994)
6. Engelbrecht, A.P.: Fundamentals of Computational Swarm Intelligence. John Wiley Sons Ltd., England (2005)
7. Guo, B., Liang, X., Wang, B., Wan, L.: Sigmoid surface control for mini underwater vehicles by improved particle swarm optimization. In: Proc. of International Conference on Robotics and Biomimetics (December 2007)
8. Kennedy, J., Eberhart, R.: Particle Swarm Optimization. In: Proc. of IEEE International Conference on Neural Networks, pp. 1942–1948. IEEE Computer Society Press, Los Alamitos (1995)
9. Kim, M.S., Kim, C.-H., Lee, J.j.: Evolving compact and interpretable takagi-sugeno fuzzy models with a new encoding scheme. IEEE Transactions on Systems, Man, and Cybernetics, Part B 36, 1006–1023 (2006)
10. Krone, A., Slawinski, T.: Data-based extraction of unidimensional fuzzy sets for fuzzy rule generation. In: Proc. of IEEE International Conference on Fuzzy Systems, vol. 02, pp. 1032–1037 (1998)
11. Nedjah, N., Mourelle, L.M.: Swarm Intelligent Systems. Springer, Heidelberg (2006)
12. Rivas, V.M., Merelo, J.J., Rojas, I., Romero, G., Castillo, P.A., Carpio, J.: Evolving two-dimensional fuzzy systems. Fuzzy Sets Systems 138(2), 381–398 (2003)
13. Setnes, M., Roubos, H.: GA-fuzzy modeling and classification: complexity and performance. IEEE Transactions on Fuzzy Systems 08, 509–522 (2000)
14. Shi, Y., Eberhart, R.C.: A Modified Particle Swarm Optimizer. In: Proc. of IEEE Congress on Evolutionary Computation, pp. 69–73. IEEE Computer Society Press, Los Alamitos (1998)
15. Wang, L.X.: The WM method completed: A flexible fuzzy system approach to data mining. IEEE Transactions on Fuzzy Systems 11, 768–782 (2003)
16. Zadeh, L.A.: Fuzzy sets. Information and Control 08, 338–353 (1965)

Maintenance Optimization of Wind Turbine Systems Based on Intelligent Prediction Tools

Zhigang Tian[1,*], Yi Ding[2], and Fangfang Ding[1]

[1] Concordia Institute for Information Systems Engineering, Concordia University, Canada
tian@ciise.concordia.ca, realoveu_i@hotmail.com

[2] School of Mechatronics Engineering, University of Electronic Science and Technology of China, China
yiding1978@hotmail.com

Wind energy is an important source of renewable energy, and reliability is a critical issue for operating wind energy systems. The Canadian wind energy industry has been growing very rapidly. The installed wind energy capacity in Canada in 2008 was approximately 2,000 mega watts (MW), which is less than one percent of the total electricity. It is believed that wind energy will satisfy 20% of Canada's electricity demand by 2025, by adding 55,000MW of new generating capacity [1]. Operation and maintenance costs account for 25-30% of the wind energy generation cost. Currently, the wind turbine manufacturers and operators are gradually changing the maintenance strategy from time-based preventive maintenance to condition based maintenance (CBM) [2-5]. In this article, we review the current research status of maintenance of wind turbine systems, and discuss the applications of artificial neural networks (ANN) based health prediction tools in this field. A CBM method based on ANN health condition prediction is presented.

1 Maintenance Optimization of Wind Turbine Systems

Maintenance management for wind power generation systems aims at reducing the overall maintenance cost and improving the availability of the systems. The existing maintenance methods for wind energy systems can be classified into corrective maintenance, preventive maintenance and condition based maintenance (CBM) [6]. In corrective maintenance, maintenance activities are performed after failure occurs. There are generally multiple wind

* Corresponding author. 1515 Ste-Catherine Street West EV-7.637, Montreal, H3G 2W1, Canada. Tel.: 1-514-848-2424 ext. 7918; Fax: 1-514-848-3171.

N. Nedjah et al. (Eds.): Innovative Computing Methods, SCI 357, pp. 53–71.
springerlink.com

turbines in a wind farm, and each wind turbine consists of multiple components, such as gearboxes, main bearings and generators, as shown in Fig. 1 [7]. Corrective maintenance actions may be performed for a component after each failure, or they can be performed after multiple components have failed to maintain the multiple failed components simultaneously [1]. Mart?nez et al demonstrated there is great need for corrective maintenance optimization [5].

The preventive maintenance is classified to time-based maintenance and age-based maintenance. In time-based preventive maintenance, the maintenance activities are typically carried out based on the predetermined time interval, say every 6 months. Age-based maintenance is employed when the component reaches a pre-defined age. However, age-based maintenance is not suitable for a multiple wind turbines farm due to expensive fixed cost, which is incurred whenever a preventive maintenance is performed once a component reaches the preventive maintenance age values [8]. To further study fixed-interval preventive maintenance, Andrawus developed the delay-time approach to optimize scheduled inspection plan, and studied a 26?600kw wind farm in the case study [9].

Condition based maintenance aims at achieving reliable and cost-effective operation of engineering systems. In CBM, condition monitoring data, such as vibration data, oil analysis data and acoustic data, are collected and processed to determine the equipment health condition; Future health condition and thus the remaining useful life (RUL) of the equipment is predicted;

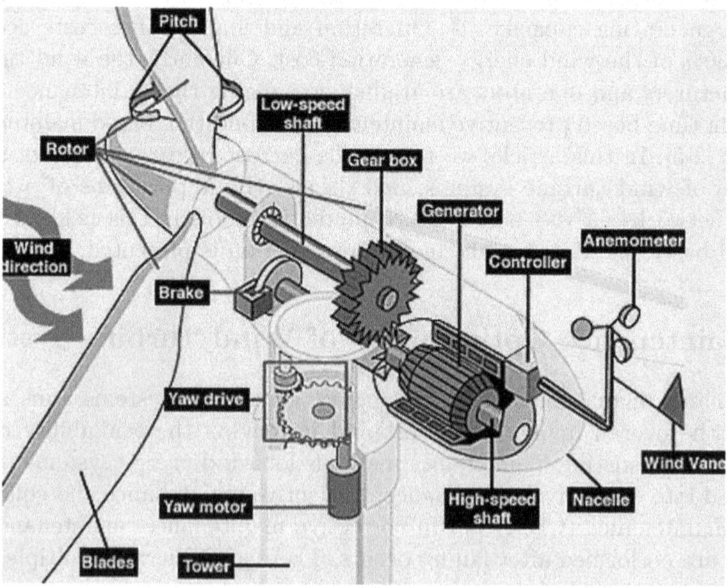

Fig. 1 Wind turbine components [7]

And optimal maintenance actions are scheduled based on the predicted future equipment health condition, so that preventive replacements can be performed to prevent unexpected failures and minimize total maintenance costs. A life cycle cost approach was adopted to evaluate the financial benefit using condition monitoring system, a tool for implementing CBM policy [10]. In [11] a multi-state Markov decision mechanism was used to estimate the wind turbine degradation process based on which the optimal maintenance scheme is devised. Tian et al. [8] developed a CBM method for wind turbine systems, based on the health condition prediction information obtained from ANN prediction models.

ANN methods have been used to investigate various problems in wind turbine systems. Yurdusev et al. proposed a method for determining the optimum tip speed ratio in wind turbines using ANN [12]. Jafarian and Ranjbar developed a combined method to estimate annual energy output of a wind turbine based on fuzzy modeling techniques and ANN [13]. Intelligent methods have also been applied to condition monitoring of wind turbines [14].

2 Critical Wind Turbine Components and Their Failure Modes

The critical components of a wind turbine system are discussed in this section. Critical wind turbine components include blades, gearboxes, main bearings and generators. Wind turbine blades are designed to collect energy from the wind and then transmit the rotational energy to the gearbox via the hub and main shaft. The number of blades and total area they cover affect wind turbine performance. Most wind turbines have only two or three blades on their rotors, the reason is that the space between blades should be great enough to avoid turbulence, so that one blade will not encounter the disturbed, weaker air flow caused by the blade which passes before it. In an offshore environment, where corrosion is a critical factor to be considered, blade material often preferred is corrosion resistant, also the possibility of achieving high strength and stiffness-to-weight ratio. Blades failures include cracks arising from fatigue, materials defects accumulating to critical cracks and ice build-up are known to cause failure.

All modern wind turbines have spherical roller bearings as main bearings. Main bearing reduces the frictional resistance between the blades, the main shaft and the gearbox while it undergoes relative motion. The main bearing is mounted in the bearing housing bolted to the main frame. Different types of wind turbines vary the quantity of bearings and bearing seats. The main bearings ensure that wind turbines withstand high loads during gusts and braking. Poor lubrication, wear, pitting, deformation of outer race and rolling elements may cause its failures.

The gearbox is one of the most important and expensive main components in the wind turbine. It is placed between main shaft and generator, the task is to increase the slow rotational speed of the rotor blades to the higher generator rotation speed. However, the gearbox in the wind turbine does not change speed just like a normal car gearbox, it always has the constant speed increasing ratio. So if a wind turbine has different operational speeds, it is because it has two different sized generators that each one has its own different rotation speed, or possibly, one generator has two different stator windings. The gearbox is connected to the generator by the coupling. The coupling is always a flexible unit made from built-in pieces of rubber, normally allowing variations of a few millimeters only. The high speed shaft from the gearbox is connected to the generator by means of a coupling. The coupling is a flexible unit made from pieces of rubber which allow some slight difference in alignment between the gearbox and the generator during normal operation. Gearbox failures include poor lubrication, bearings and gear teeth failures can cause major failures.

The generator transforms mechanical energy into electrical energy. The blades transfer the kinetic energy from the wind into rotational energy, and then generator supply the energy from the wind turbine to the electrical grid. Generator produces either alternating current (AC) or direct current (DC), and they are available in a large range of output power ratings. The generator's rating, or size, is dependent on the length of the wind turbine blades because more energy is captured by longer blades. Bearings are the major cause of failure of generator. Thus, maintenance is mainly restricted to bearing lubrication.

Acronyms

CBM: condition based maintenance,
ANN: artificial neural network.

3 Component Health Condition Prognostics Using ANN

3.1 Health Condition Prediction

The objective of health condition prognostics is to predict the equipment future health conditions and thus the remaining useful life. At each inspection point, the condition monitoring measurements are collected, and the health condition prognostics methods can be used to produce the predicted failure time value or the remaining useful life value, and some prognostics methods can give the associated prediction uncertainties as well. The health condition prediction methods can be divided into model-based methods and data-driven methods. The model-based methods, also known as the physics-of-failure methods, perform prognostics using equipment physical models and damage propagation models. Model-based prognostics methods have been reported for components such as bearings (Marble et al [15]) and gearboxes (Kacprzynski et al [16], Li and Lee [17]). The key limitation of the model-based methods is that for some components and systems, authentic physics-of-failure models are very difficult to build because equipment damage propagation processes and dynamic responses are very complex. Data-driven methods directly utilize the collected condition monitoring data for health condition prediction, and do not require physics-of-failure models. Examples of the data-driven methods include the proportional hazards model method developed by Banjevic et al [18], the Bayesian prognostics methods [19], and the ANN based prognostics methods [20-22].

Outputs of the prognostics methods are predicted failure time and the associated uncertainty. That is, at a certain inspection point, the predicted failure time distribution can be obtained for the component being monitored. Among various data-driven methods, ANN based methods have been considered to be very effective and flexible tools for component health condition prognostics. In this work, we use the ANN prediction approach developed in [21]. The ANN model used in this approach is shown in Fig. 2, which is a feedforward neural network model with one input layer, two hidden layers and one output layer. The inputs of the ANN are the component age values and the condition monitoring measurements at the current and previous inspection points. In the example of the ANN model shown in Fig. 2, there are two condition monitoring measurements. Specifically, t_i is the age of the component at the current inspection point i, and t_{i-1} is the age at the previous inspection point $i-1$. $z_i^1$and $z_{i-1}^1$are values of measurement 1 at the current and previous inspection points, and $z_i^2$and z_{i-1}^2 are values of measurement 2 at the current and previous inspection points. The output of the ANN model is the life percentage at current inspection time, denoted byP_i. For example,

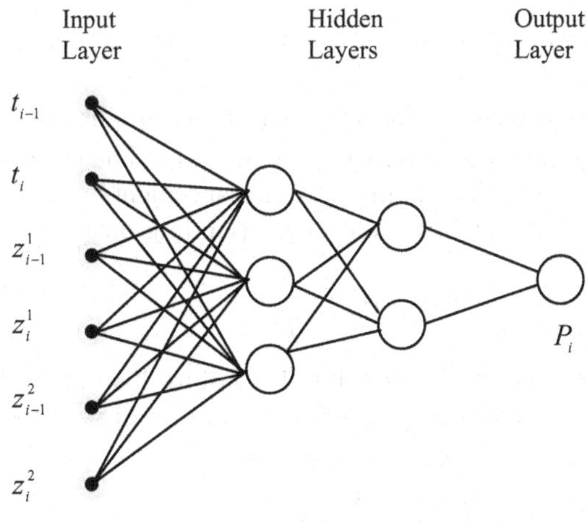

Fig. 2 Structure of the ANN model for component health condition prediction [21]

if the failure time of a component is 850 days and the age of the component at the current inspection point is 500 days, the life percentage value would be $P_i = 500/850 \times 100\% = 58.82\%$.

The ANN model utilizes suspension histories as well as failure histories. A failure history of a unit refers to the period from the beginning of the component life to the end of its life, a failure, and the inspection data collected during this period. In a suspension history, though, the unit is taken out of service before the failure occurs. Once trained, the ANN prediction model can be used to predict the remaining life based on the age value of the component and the condition monitoring measurements. As mentioned above, the output of the ANN model is life percentage, based on which the predicted failure time can be calculated. For example, at a certain inspection point, if the age of the component is 400 days and the life percentage predicted using ANN is 80%, the predicted failure time will be 400/80% = 500 days.

3.2 An ANN Health Condition Prediction Method

Now we present the ANN health condition prediction method developed by Tian et. al [21]. The procedure of the method is shown in Fig. 3. The detailed explanations of the procedure are given in the remainder of this section.

3.2.1 Constructing the Failure History Training Data Set

The first step of the approach is to construct the failure history training data set, which will be combined with training data set based on the suspension histories to train the ANN. Suppose there are J condition monitoring measurements used in the ANN model. An ANN input vector based on failure history f takes the following form:

$$\mathbf{I}N = \left(t_{f,\, i-1},\, t_{f,\, i},\, z^1_{f,\, i-1},\, z^1_{f,\, i},\, z^2_{f,\, i-1},\, z^2_{f,\, i},\, ?,\, z^J_{f,\, i-1},\, z^J_{f,\, i}\right), \quad (1)$$

where $t_{f,i}$ denotes the equipment age at inspection point i in failure history f, and $z^j_{f,\, i}$ represents the measurement j at time $t_{f,i}$. The input vector contains the time and the condition monitoring measurements values at the current and previous inspection points. The corresponding output value is:

$$P_{f,\, i} = \frac{t_{f,i}}{TF_f}, \quad (2)$$

where TF_f represents the failure time for failure history f. Thus, the total number of input/output pairs based on the failure histories is:

$$N_F = \sum_{f=1}^{F} (NF_f - 1), \quad (3)$$

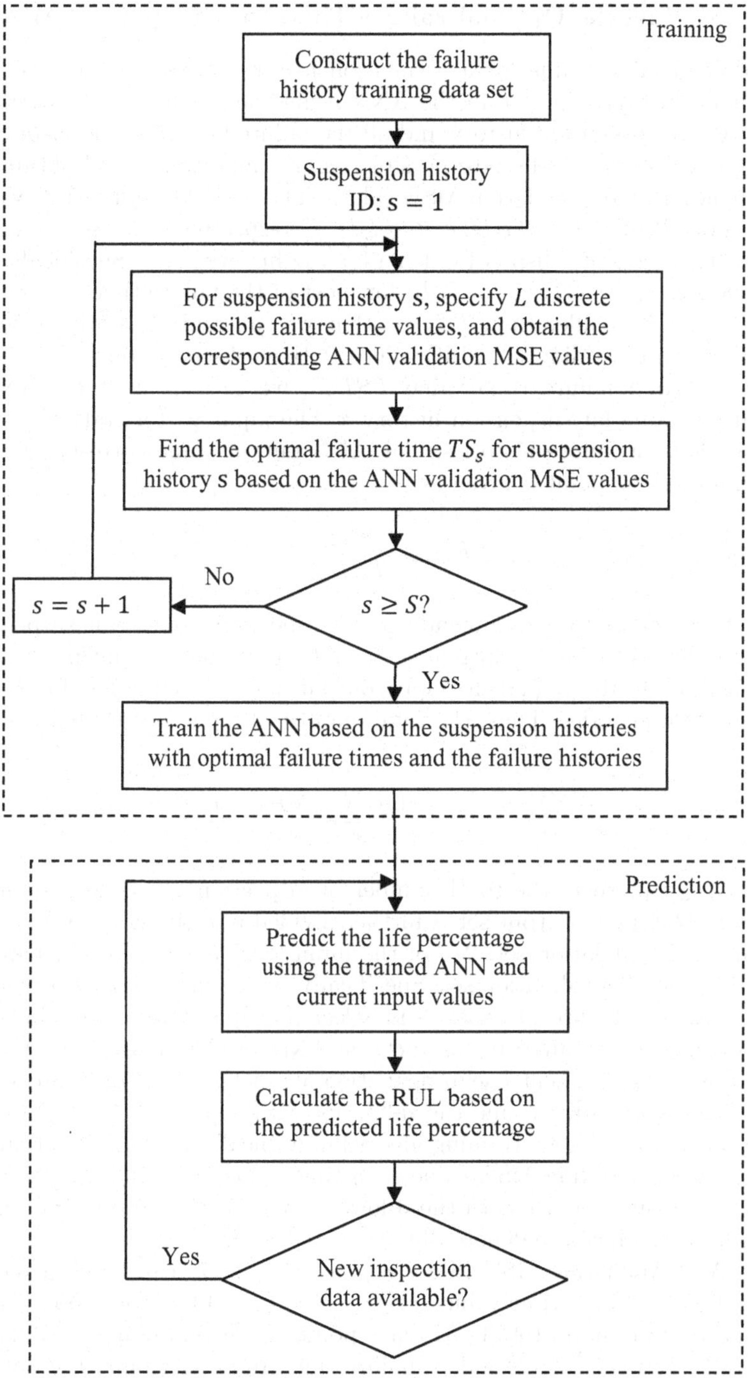

Fig. 3 Procedure of the remaining life prediction approach [21]

3.2.2 Finding the Optimal Failure Time for a Suspension History

The optimal failure time for a suspension history corresponds to the lowest validation MSE if we train the ANN using the training set constructed based on this suspension history and all the failure histories. For suspension history s, we specify L discrete possible failure time values, and obtain the corresponding ANN validation MSE values. The discrete failure time values are denoted by $TSD_{s,1}$, $TSD_{s,2}$, ?, $TSD_{s,L}$, respectively. These values are selected based on the suspension time for the history, TS_s. Specifically, we can have $TSD_{s,l} \geq TS_s$ $(1 \leq l \leq L)$ for most of the failure time values, and have 1-2 values smaller than TS_s, so that we can find the optimal failure time based on the validation MSE values at these discrete points.

For a certain failure time value $TSD_{s,l}$, we can obtain the ANN input/output pairs for suspension history s. The input vectors take the same form as that for failure histories. The ANN output value corresponding to the ith inspection point is given as:

$$P_{s,\ i,l} = \frac{t_{s,i}}{TSD_{s,l}}, \tag{4}$$

where $t_{s,i}$ denotes the equipment age at inspection point i in suspension history s. The ANN input/output set includes the input/output pairs based on suspension history s and the input/output data set constructed based on all the failure histories. Thus, the total number of input/output pairs is:

$$N_{Ss} = NS_s - 1 + \sum_{f=1}^{F} (NF_f - 1), \tag{5}$$

where NS_s represents the total number of inspection points in suspension history s. The input/output set is further divided into the ANN training set and the ANN validation set: 2/3 of the input/output pairs for the training set and 1/3 for the validation set. Specifically, we go through the suspension history and the failure histories, and select an input/output pair in every three input/output pairs to construct the ANN validation set. The ANN is trained using the resilient backpropagation algorithm based on the training set and the validation set, and the validation MSE can be obtained. Because of the randomness in the training algorithm, typically we cannot obtain the exactly same validation MSE value each time. Thus, in this work, we train the ANN 30 times, and record the 3 lowest, or best, validation MSE values for future use, which are denoted by $ve^s_{l,r}$ $(r = 1,\ 2,\ 3)$.

The ANN validation MSE values $ve^s_{l,r}$ $(r = 1,\ 2,\ 3)$ can be obtained for all the discrete failure time values $TSD_{s,l}$ $(1 \leq l \leq L)$ for suspension history s. Thus, we can obtain totally $3L$ data points, each containing a validation MSE value and the corresponding failure time. In order to find the optimal failure time based on the discrete validation MSE values, we need to fit these validation MSE data points. Considering the flexibility and the ability to model simple trends, we use the third order polynomial to fit the data points:

$$y = ax^3 + bx^2 + cx + d, \tag{6}$$

where y represents the validation MSE, x represents failure time, and a, b, c, d represent the polynomial coefficients to be determined. The objective of using the polynomial function in Equation (6) is to build a continuous function to represent the change in the validation MSE with respect to the possible failure time. Once the polynomial function is obtained, it is easy to find the optimal failure time corresponding to the lowest ANN validation MSE, using a simple optimization process. The optimal failure time for suspension history s is denoted by TS_s^*. To enforce the suspension time constraint, let $TS_s^* = TS_s$ if TS_s^* is smaller than the suspension time.

3.2.3 ANN Training Based on the Suspension Histories with Optimal Failure Times and the Failure Histories

Using the procedure in Section 3.3, we can find the optimal failure times for all the suspension histories: TS_s^* $(1 \leq s \leq S)$. Now we can train the ANN for remaining useful prediction based on the suspension histories with optimal failure times and the failure histories. The ANN output value for an input/output pair from a failure history is given by Equation (4), and that from a suspension history is given as follows:

$$P_{s,\ i} = \frac{t_{s,i}}{TS_s^*} \tag{7}$$

Thus, the total number of input/output pairs is:

$$N_{\text{IO}} = \sum_{s=1}^{S} (NS_s - 1) + \sum_{f=1}^{F} (NF_f - 1), \tag{8}$$

The ANN training set includes 2/3 of the input/output pairs, and the ANN validation set includes the remaining 1/3 of the input/output pairs. Similarly, we train the ANN 30 times using the resilient backpropagation algorithm, and save the ANN with the smallest validation MSE.

3.2.4 Remaining Life Prediction Using Trained ANN

Once the ANN is trained, as discussed in the previous section, it can be used for RUL prediction for other equipments being monitored. The age and condition monitoring measurements at the current and previous data points are used as inputs to the trained ANN, and the current life percentage can be obtained. The RUL is obtained by dividing the current age by the predicted life percentage. When new condition monitoring data is available, the prediction will be performed again and the RUL will be updated. The remaining useful life prediction process stops when the equipment fails or when it is preventively taken out of service.

3.3 Quantification of the ANN Health Condition Prediction Uncertainty

In this section, we present a method for estimating the predicted failure time distribution based on the ANN life percentage prediction errors obtained during the ANN training and testing processes [23].

In the ANN training process, the ANN model is trained based on the available failure histories and suspension histories. The ANN model inputs include the age data and the condition monitoring measurements at the current and previous inspection points. The output of the ANN model is the life percentage of the inspected component at the current inspection point, denoted by P_i. In the training process, the weights and the bias values of the ANN model are adjusted to minimize the error between the ANN output and the actual life percentage. After ANN training is completed, the prediction performance of the trained ANN model is tested using testing histories which are not used in the training process. Here, the ANN prediction error is defined as the difference between the ANN life percentage output and the actual life percentage value at an inspection point in the test histories. That is, the ANN prediction error at inspection point k in a test history is equal to $(\hat{P}_k$-$P_k)$, where P_k denotes the actual life percentage and $\hat{P}_k$ is the predicted life percentage using ANN. Since a test history contains many inspection points, with several test histories, we can obtain a set of ANN life percentage prediction error values.

In this study, it is assumed that the ANN life percentage prediction error is normally distributed, since the prediction uncertainty is mainly due to the capability of the ANN prediction model. With the obtained set of ANN prediction error values, we can estimate the mean $\mu_p$and standard deviationσ_p of the ANN life percentage prediction error. Suppose at a certain inspection point, the ANN life percentage output is P_t, then the mean of the predicted life percentage should be $P_t - \mu_p$, and the standard deviation is still σ_p. If the age of the component at the current inspection point is t, the predicted failure time will be $t/\left(P_t - \mu_p\right)$, and the standard deviation of the predicted failure time will be $\sigma_p \cdot t/\left(P_t - \mu_p\right)$. That is, the predicted failure time T_p follows the following normal distribution:

$$T_p \sim N\left(t/\left(P_t - \mu_p\right),\ \sigma_p \cdot t/\left(P_t - \mu_p\right)\right). \tag{9}$$

4 A CBM Approach for Wind Power Generation Systems

In this section, the CBM method developed by Tian et al. [8] for wind power generation systems is presented. Suppose there are N wind turbines in the wind farm, and we consider M critical components for each turbine. In this work, it is assumed that all the wind turbines under consideration are identical, and the degradation processes of the wind turbine components are mutually independent.

4.1 Failure Probability Estimation at the Component and Turbine Levels

At the wind turbine component level, condition monitoring data, such as vibration data and acoustic emission data, can be collected, and failure time distribution can be predicted for each component using the prognostics methods presented in Section 2. It is assumed that the predicted failure time follows the normal distribution, as discussed in Section 2. The failure probabilities for the wind turbine components, which will be defined later, can be calculated based on the predicted failure time distributions, and the CBM decisions will be made based on the failure probabilities. The failure probability for component m in turbine n is defined as follows [23]:

$$\Pr_{n,m} = \frac{\int_t^{t+L} \frac{1}{\sigma\sqrt{2\pi}} e^{-\frac{1}{2}\left(\frac{x-t_p}{\sigma}\right)^2} dx}{\int_t^{\infty} \frac{1}{\sigma\sqrt{2\pi}} e^{-\frac{1}{2}\left(\frac{x-t_p}{\sigma}\right)^2} dx} \tag{10}$$

where t is the age of the component at the current inspection point, t_p is the predicted failure time using ANN, and σ is the standard deviation of the predicted failure time distribution. Based on the discussions in Section 2, we have the following relationships:

$$t_p = t/\left(P_t - \mu_p\right), \sigma = \sigma_p \cdot t/\left(P_t - \mu_p\right). \tag{11}$$

L in Equation (10) is the maintenance lead time, which is defined as the interval between the time maintenance decision is made and the time when the maintenance actions are performed. The lead time consists of the time required to assemble the maintenance team, order the replacement parts, prepare the maintenance equipments to perform the maintenance, and travel to the wind farm, etc. Thus, the maintenance decisions made at the current inspection point can affect the wind turbines only when the lead time has passed, and we have no influence on the failures during the lead time. So, it is reasonable to base the maintenance decisions on the failure probabilities during the maintenance lead time L in order to reduce the failure risks. To reasonably simplify the problem, we assume L is the same for all maintenance actions in this studyto simplify our discussion.

If we focus on the critical components in a turbine, such as rotor, main bearing, gearbox, generator, etc., the turbine can be treated as a series system. That is, the failure of any turbine component will cause the failure of the turbine. Thus, the failure probability of wind turbine n during lead time L can be calculated as follows:

$$\Pr_n = 1 - \prod_{m=1}^{M} \left(1 - \Pr_{n,m}\right) \quad (12)$$

4.2 The Proposed CBM Policy

For the purpose of simplifying the descriptions, we use replacement to refer to a maintenance action, such as the replacement of the main bearing, or the replacement of a faulty gear within the gearbox. Suppose wind turbine components are continuously monitored. Maintenance decisions are made based on the failure probabilities of the components and the wind turbines, which can be calculated based on the component health condition monitoring and prognostics information.

The proposed CBM policy for the wind power generation systems is summarized as follows:

1. Perform failure replacement if a component fails. The maintenance equipments and replacement parts will be scheduled, and the maintenance team will be sent to the wind farm.
2. Send a maintenance team to the wind farm and perform preventive replacements if any wind turbine in the wind farm is determined to be maintained.
3. Perform preventive replacements on components in wind turbine n if $Pr_n > d_1$, where Pr_n is the failure probability of the wind turbine n, and d_1 is the pre-specified level 1 failure probability threshold value at the turbine level.
4. If turbine n is to be performed preventive replacement on, perform preventive replacement on its components in order to bring the turbine failure probability down to below d_2. d_2 is called the level 2 failure probability threshold value at the turbine level.

As can be seen, once the two failure probability threshold values, d_1 and d_2, are specified, the CBM policy is determined.

4.3 CBM Optimization Model and Solution Method

Based on the above CBM policy, the CBM optimization model can be briefly formulated as follows:

$$\begin{aligned} &\min \quad C_E\left(d_1, d_2\right) \\ &\text{s.t.} \\ &\qquad 0 < d_2 < d_1 < 1 \end{aligned} \tag{13}$$

where C_E is the total expected maintenance cost per unit of time for a certain CBM policy defined by the two failure probability threshold values d_1 and d_2. d_1 and d_2 take real values between 0 and 1, and $d_2 < d_1$. The objective of the CBM optimization to find the optimal d_1 and d_2 values to minimize the total maintenance cost. The optimization functions built in Matlab can be used to solve this optimization problem, and find the optimal threshold failure probability values.

Before the optimization can be performed, we need to first be able to calculate the cost value C_E given two failure probability threshold values d_1 and d_2. Due to the complexity of the problem, it is very difficult to develop a numerical algorithm for the cost evaluation of the CBM policy for the wind power generation systems. In this paper, we present a simulation method for the cost evaluation. The flow chart for the procedure of the simulation method is presented in Fig. 4, and detailed explanations of the procedure are given in the following paragraphs.

Step 1: Building the ANN prediction model. For each type of wind turbine component, determine the life time distribution based on the available failure and suspension data. Weibull distributions are assumed to be appropriate for components lifetime, and the distribution parameters α_m and β_m can be estimated for each component m. For each type of component, based on the available failure and suspension histories, an ANN prediction model can be trained, and the mean and standard deviation of the ANN life percentage prediction error, which are denoted by $\mu_{p,m}$ and $\sigma_{p,m}$, respectively, can be calculated.

Step 2: Simulation initialization. As mentioned earlier, suppose there are N wind turbines considered in the wind farm, and M critical components are considered for each turbine. Specify the maximum simulation time T_{Max}, and the inspection interval T_I. T_I can be set to be a small value, say 1 day, so that we can approximately achieve continuous monitoring. Or we can set T_I to be a bigger value, say 10 days, to improve computation efficiency and achieve reasonably accurate results. For each component m, specify the cost values, including the failure replacement cost $c_{f,m}$ and the variable preventive replacement cost $c_{p,m}$. The fixed cost of maintaining a certain wind turbine, $c_{p,T}$, and the fixed cost of sending a maintenance team to the wind farm, c_{Farm}, also need to be specified. The total replacement cost is set to be $C_T = 0$, and will be updated during the simulation process. At time set $t_{ABS} = 0$, generate the real failure times for each component in each turbine. That is, for component m in turbine n, generate a real failure time $TL_{n,m}$ by sampling the Weibull distribution for component m with parameters α_m and β_m. Thus, at time 0, the age values for all the components are 0.

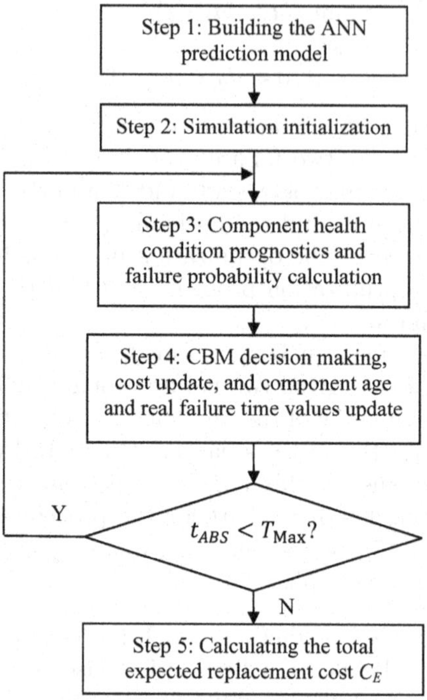

Fig. 4 Flow chart for the proposed simulation method for cost evaluation

Step 3: Component health condition prognostics and failure probability calculation. At a certain inspection point when the time is t_{ABS}>0, the age of component m in turbine n is represented by $t_{n,m}$, and its real failure time is known at this point, which is $TL_{n,m}$. For each component, generate the predicted failure time, $TP_{n,m}$, by sampling the normal distribution $N\left(TL_{n,m}, \sigma_p \cdot TL_{n,m}\right)$. Based on the discussion in Section 2, the predicted failure time distribution can be obtained as $N\left(TP_{n,m}, \sigma_p \cdot TP_{n,m}\right)$. Thus, the current failure probability during the lead time for the component is:

$$\Pr_{n,m} = \frac{\int_{t_{n,m}}^{t_{n,m}+L} \frac{1}{\sigma_p TP_{n,m}\sqrt{2\pi}} e^{-\frac{1}{2}\left(\frac{x-TP_{n,m}}{\sigma_p TP_{n,m}}\right)^2} dx}{\int_{t_{n,m}}^{\infty} \frac{1}{\sigma_p TP_{n,m}\sqrt{2\pi}} e^{-\frac{1}{2}\left(\frac{x-TP_{n,m}}{\sigma_p TP_{n,m}}\right)^2} dx} \tag{14}$$

The failure probabilities of each turbine can be calculated using Equation (12) based on the failure probabilities of its components.

Step 4: CBM decision making and cost update. At the current inspection point t_{ABS}, the CBM decisions can be made according to the CBM policy, described in Section 3.2, based on the failure probabilities of the turbines and their components. If the current time t_{ABS} has not exceeded the maximum simulation time T_{Max}, repeat Step 3 and Step 4.

Step 5: Total replacement cost calculation. When the maximum simulation time is reached, that is, $t_{ABS} = T_{\text{Max}}$, the simulation process is completed. The total replacement cost for the wind farm can be calculated as:

$$C_E = C_T / T_{\text{Max}}. \tag{15}$$

And the total replacement cost for each turbine is:

$$C_{ET} = \frac{C_T}{N \cdot T_{\text{Max}}}. \tag{16}$$

5 An Example

In this section, an example is used to demonstrate the proposed CBM approach for wind power generation systems [8]. Consider a group of 5 wind turbines, produced and maintained by a certain manufacturer, in a wind farm at a remote site. To simplify our discussion, in this example, we study 4 key components in each wind turbine: the rotor (including the blades), the main bearing, the gearbox and the generator, as shown in Fig. 5 [24].

Assume the Weibull distributions are appropriate to describe the component failure times, and the Weibull parameters are given in Table 1. The component lifetime distribution parameters are specified based on the data given in Ref. [25] and [26]. The cost data are given in Table 2, including the failure replacement costs for the components, the fixed and variable

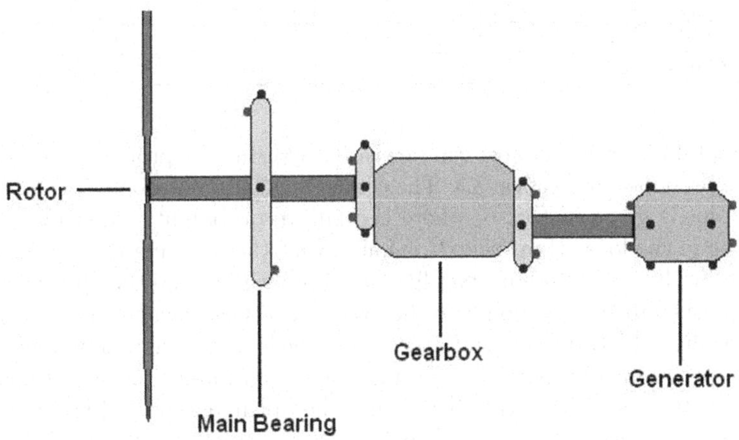

Fig. 5 Key wind turbine components considered in the example [24]

Table 1 Weibull failure time distribution parameters for major components

Component	Scale parameter α(days)	Shape parameter β
Rotor	3,000	3.0
Main bearing	3,750	2.0
Gearbox	2,400	3.0
Generator	3,300	2

preventive replacement costs and the cost of sending a maintenance team to the wind farm. The cost data are specified based on the cost related data given in Ref. [1] and [27]. The ANN prediction method can be used to predict the failure time distributions of the wind turbine components, and suppose the standard deviations of the ANN life percentage prediction errors are 0.12, 0.10, 0.10, and 0.12, respectively, as shown in Table 3. The standard deviation values are selected by referring to that estimated using the bearing degradation data in Ref [18] and [28]. The maintenance lead time is assumed to be 30 days, and the inspection interval is set at 10 days.

Table 2 Failure replacement and preventive maintenance costs for major components

Component	Failure replacement cost ($1000)	Variable preventive maintenance cost ($1000)	Fixed preventive maintenance cost ($1000)	Fixed cost to the wind farm ($1000)
Rotor	112	28		50
Main bearing	60	15	25	
Gearbox	152	38		
Generator	100	25		

The total maintenance cost can be evaluated using the proposed simulation method presented in Section 3.3. The cost versus failure probability threshold values plot is given in Fig. 6, where the failure probability threshold values are given in the logarithm scale. It is found that the total maintenance cost is affected by the two failure probability threshold values, and the optimal CBM policy exists which corresponds to the lowest cost. Optimization is performed, and the optimal CBM policy with respect to the lowest total maintenance cost can be obtained. The obtained optimal threshold failure probability values are: $d_1 = 0.1585$, $d_2 = 3.4145?10^{-6}$, and the optimal expected maintenance cost per unit of time is 577.08 $/day.

Table 3 ANN life percentage prediction error standard deviation values for major components

Component	Standard deviation
Rotor	0.12
Main bearing	0.10
Gearbox	0.12
Generator	0.10

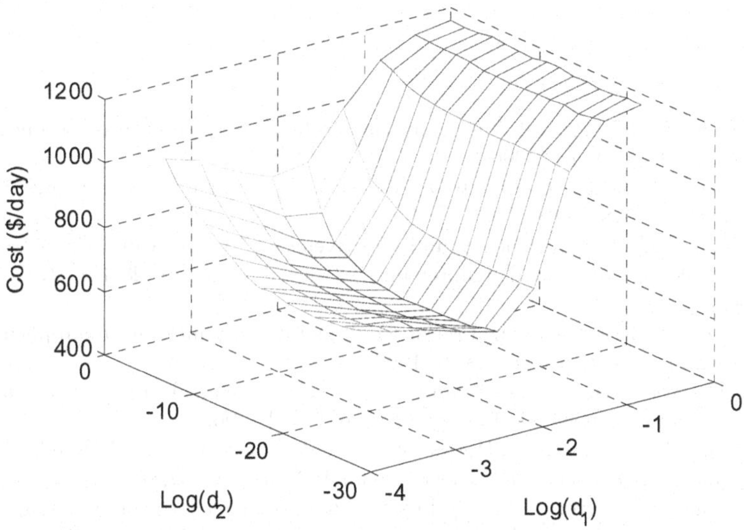

Fig. 6 Cost versus failure probability threshold values in the logarithm scale

6 Conclusions

Wind energy is an important source of renewable energy, and reliability is a critical issue for operating wind energy systems. Maintenance optimization approaches aiming at improving wind turbine system reliability and reducing the overall operating costs. Currently, the wind turbine manufacturers and operators are gradually changing the maintenance strategy from time-based preventive maintenance to condition based maintenance. Given the failure probabilities for components and the system, optimal CBM decisions can be made on target wind turbines to be maintained, the maintenance schedule, and key components to be inspected and fixed. Future research includes developing wind turbine component health monitoring methods by utilizing data collected under different conditions, building more accurate wind turbine system reliability models considering the degraded system performance, and maintenance optimization for farms with heterogeneous wind turbines, etc.

Acknowledgments

This research is supported by the Natural Sciences and Engineering Research Council of Canada (NSERC) and Le Fonds qu?b?cois de la recherche sur la nature et les technologies (FQRNT).

References

1. Hau, E.: Wind turbines: fundamentals, technologies, application, economics. Springer, Heidelberg (2006)
2. Tavner, P.J., Xiang, J., Spinato, F.: Reliability analysis for wind turbines. Wind Energy 10, 1–18 (2007)
3. Krokoszinski, H.J.: Efficiency and effectiveness of wind farms - keys to cost optimized operation and maintenance. Renewable Energy 28(14), 2165–2178 (2003)
4. Sen, Z.: Statistical investigation of wind energy reliability and its application. Renewable Energy 10(1), 71–79 (1997)
5. Martinez, E., Sanz, F., Pellegrini, S.: Life cycle assessment of a multi-megawatt wind turbine. Renewable Energy 34(3), 667–673 (2009)
6. Jardine, A.K.S., Tsang, A.H.: Maintenance, Replacement, and Reliability Theory and Applications. CRC Press, Taylor & Francis Group (2006)
7. Sterzinger, G., Svrcek, M.: Wind turbine development: location of manufacturing activity. Renewable Energy Policy Project Technical Report (2004)
8. Tian, Z., Jin, T., Wu, B., Ding, F.: Condition based maintenance optimization for wind power generation systems under continuous monitoring. Renewable Energy (2010) (accepted)
9. Andrawus, J.A.: Maintenance optimization for wind turbines Ph.D thesis, The Robert Gordon University (2008)
10. Nilsson, J., Bertling, L.: Maintenance management of wind power systems using condition monitoring systems—life cycle cost analysis for two case studies. IEEE Transactions on Energy Conversion 22(1), 223–229 (2007)
11. Byon, E., Ding, Y.: Season-dependent condition based maintenance for a wind turbine using a partially observed Markov decision process. IEEE Transactions on Power Systems (2010) (accepted)
12. Yurdusev, M.A., Ata, R., Cetin, N.S.: Assessment of optimum tip speed ratio in wind turbines using artificial neural networks. Energy 31(12), 2153–2161 (2006)
13. Jafarian, M., Ranjbar, A.M.: Fuzzy modeling techniques and artificial neural networks to estimate annual energy output of a wind turbine. Renewable Energy 35(9), 2008–2014 (2010)
14. Hameed, Z., Hong, Y.S., Cho, Y.M., Ahn, S.H., Song, C.K.: Condition monitoring and fault detection of wind turbines and related algorithms: A review. Renewable and Sustainable Energy Reviews 13(1), 1–39 (2009)

15. Marble, S., Morton, B.: Predicting the remaining life of propulsion system bearings. In: Proceedings of the 2006 IEEE Aerospace Conference, Big Sky, MT, USA (2006)
16. Kacprzynski, G.J., Roemer, M.J., Modgil, G., Palladino, A., Maynard, K.: Enhancement of physics-of-failure prognostic models with system level features. In: Proceedings of the 2002 IEEE Aerospace Conference, Big Sky, MT, USA (2002)
17. Li, C.J., Lee, H.: Gear fatigue crack prognosis using embedded model, gear dynamic model and fracture mechanics. Mechanical Systems and Signal Processing 19, 836–846 (2005)
18. Banjevic, D., Jardine, A.K.S., Makis, V., Ennis, M.: A control-Limit ploicy and software for condition based maintenance optimization. Infor. 39, 32–50 (2001)
19. Gebraeel, N., Lawley, M., Li, R., Ryan, J.K.: Life Distributions from Component Degradation Signals: A Bayesian Approach. IIE Transactions on Quality and Reliability Engineering 37(6), 543–557 (2005)
20. Tian, Z.: An artificial neural network method for remaining useful life prediction of equipment subject to condition monitoring. Journal of Intelligent Manufacturing (2009) (accepted), doi: 10.1007/s10845-009-0356-9
21. Tian, Z., Wong, L., Safaei, N.: A neural network approach for remaining useful life prediction utilizing both failure and suspension histories. Mechanical Systems and Signal Processing 24(5), 1542–1555 (2010)
22. Tian, Z., Zuo, M.: Health condition prediction of gears using a recurrent neural network approach. IEEE Transactions on Reliability (2010) (accepted)
23. Wu, B., Tian, Z., Chen, M.: Condition based maintenance optimization using neural network based health condition prediction. In: IEEE Transactions on Systems, Man, and Cybernetics–Part A: Systems and Humans (2010) (under review)
24. National instruments products for wind turbine condition monitoring (2010), `http://zone.ni.com/devzone/cda/tut/p/id/7676`
25. 2008-2009, WindStats Newsletter, 21(4)-22(3), Denmark
26. Guo, H.T., Watson, S., Tavner, P.: Reliability analysis for wind turbines with incomplete failure data collected from after the date of initial installation. Reliability Engineering and System Safety 94(6), 1057–1063 (2009)
27. Fingersh, L., Hand, M., Laxson, A.: Wind turbine design cost and scaling model. National Renewable Energy Laboratory technical report (2006)
28. Stevens, B.: EXAKT reduces failures at Canadian Kraft Mill, (2006), `http://www.modec.com`

Clonal Selection Algorithm Applied to Economic Dispatch Optimization of Electrical Energy

Daniel Cavalcanti Jeronymo[1], Leandro dos Santos Coelho[1,2], and Yuri Cassio Campbell Borges[3]

[1] Department of Electrical Engineering, PPGEE, Federal University of Parana, UFPR, Polytechnic Center, Zip code 81531-970, Curitiba, Parana, Brazil
dcavalcanti@bigfoot.com

[2] Industrial and Systems Engineering Graduate Program, PPGEPS, Mechanical Engineering Graduate Program, PPGEM, Pontifical Catholic University of Parana, PUCPR, Imaculada Conceicao, 1155, Zip code 80215-901, Curitiba, PR, Brazil
leandro.coelho@pucpr.br

[3] Telecommunications and Control Department, Polytechnique School, University of Sao Paulo (USP), Professor Luciano Gualberto, Travessa 3, 158, CEP 05508-900, Sao Paulo, Sao Paulo, Brazil
yuri.campbell@gmail.com

Economic dispatch is an important problem in power systems. This chapter presents how a method of stochastic optimization, a metaheuristic known as CLONALG (CLONal selection ALGorithm), can be applied to the economic dispatch problem. The objective function used in the optimization is based on Karush-Kuhn-Tucker conditions, thus, guaranteeing a convergence to the global optimum. Examples and results are presented showing the method is capable of finding the optimum solution while respecting power generation limits.

1 Fundamentals of CLONal Selection ALGorithm (CLONALG)

Clonal selection theory [5] is an widely accepted model for acquired immunity. According to the theory, the continuous encounter of lymphocytes and antigens causes biological triggers that lead to a selection process of the lymphocytes, causing the antibody population to have a greater affinity for that antigen. This process is, effectively, a minimization of the error on the function of antibody encounters with antigens. This theory gave origin, in

N. Nedjah et al. (Eds.): Innovative Computing Methods, SCI 357, pp. 73–83.
springerlink.com

computing, to the class of clonal selection algorithms of immune artificial systems (IAS).

The theory of clonal selection has the following characteristics: maintenance of a memory set, selection and cloning of the most activated antibodies, death of non-activated antibodies, affinity maturation through the mutation process, reselection of clones proportional to the antigen affinity, generation and maintenance of diversity.

Aiming to create a minimization or classification algorithm with similar characteristics, the clonal selection algorithm (CSA) was proposed in [6] and later, with a more developed concept, the new concept of the algorithm was published by [7] and named clonal selection algorithm (CLONALG). The pseudo code of the algorithm is detailed in Table 1.

Table 1 Pseudo code of CLONALG

$[f^*, f, A_b] = CLONALG(F_{obj}, N_{abs}, N_{sel}, N_{new}, \beta, \rho, N_{gens})$	
$[A_b] = generate(N_{abs});$	{Generate initial population}
$for\ i = 1 : N_{gens}\ do$	
$[f] = evaluate(F_{obj}, A_b);$	{Determine population affinity}
$[A_{b_n}, f_n] = select(A_b, f, N_{sel});$	{Select best}
$[C, f_c] = clone(A_{b_n}, f_n, \beta);$	{Proportional cloning}
$[C_{mut}] = hypermutate(C, f_c, \rho);$	{Affinity maturation}
$[f_{mut}] = evaluate(F_{obj}, C_{mut});$	{Determine clone affinity}
$[A_{b_n}, f_n] = select(C_{mut}, f_{mut}, N_{sel});$	{Select best}
$[A_b, f] = insert(A_b, A_{b_n}, f, f_n);$	{Insert clones}
$[A_{b_d}] = generate(N_{new});$	{Generate new antibodies}
$[A_b] = replace(A_b, A_{b_d}, f);$	{Replace worst}
$end\ for$	
$[f] = evaluate(F_{obj}, A_b);$	{Determine final solutions}
$[f^*] = min(f);$	{Determine best solution}

The algorithm in it's original version presents affinity proportional cloning, similar to the biological inspiration. This is achieved by sorting the antibody population by affinity to the antigen in ascending order and then calculating the number of clones for each antibody according to:

$$N_{clones} = round\left(\frac{\beta \cdot N_{abs}}{i} + 0.5\right) , \tag{1}$$

where i is the antibody rank, $i \in [1, N_{abs}]$. However, in the optimization version of CLONALG the affinity proportional cloning is not effective [10]. The recommended approach is to use the equation given by [7]:

$$N_{clones} = round\left(\beta \cdot N_{abs}\right) . \tag{2}$$

The mutation, also known as affinity maturation, is given with a rate inversely proportional to the fitness of each antibody. That is, antibodies with a lower affinity to the antigen will suffer more mutations while the antibodies with a better affinity will suffer less mutations. This is done by first normalizing the objective function of each antibody, f_i, in the range $[0, 1]$. Then a random perturbation vector is generated for each clone created from the antibodies, this vector is scaled by the mutation rate, which is given by the following equation:

$$\alpha = e^{(-\rho \cdot f_i)} . \tag{3}$$

1.1 Parameters

CLONALG's performance depends heavily on the user defined parameters, which are: antibody population size, selection pool size, remainder replacement pool size, clonal factor and mutation factor. Basic output parameters are: F_{obj} , the objective function to be minimized; f^*, the best objective function value found; f, a vector of final values for the objective function; and A_b, a matrix of final solutions.

Antibody population size (N_{abs}) — The total amount of antibodies to be used by the algorithm. Antibodies are an analogy to solution vectors for an objective function.

Selection pool size (N_{sel}) — The number of best antibodies to be selected by the algorithm for the cloning stage. This is similar to an elitism approach, like the one used in Genetic Algorithms (GA). Only clones and new antibodies are passed on to newer generations, thus, the value of N_{sel} determines the selective pressure on the population. Greater values soften the selective pressure, ensuring more members of the population are cloned and the search is broadened across the search space, causing the population diversity to increase. On the other hand, lower values can increase the selective pressure by allowing only a few solutions with greater affinity to be cloned, causing a portion of the antibody population to be dominated by these solutions, reducing diversity. The trade-off between exploitation, local search, and exploration, global search, is mainly done by the selection of this parameter.

Remainder replacement pool size (N_{new}) — The number of lower affinity antibodies to be replaced by random antibodies on each generation. Essentially, this is a mechanism to ensure a minimum, as in additional, diversity is maintained in the antibody population. In practice, this mechanism is only useful when the algorithm is stuck on local optima. This additional diversity can be disabled by setting the value to zero.

Clonal factor (β) — A scaling factor for the number of clones created for each selected antibody. Common values are $\beta \in (0, 1]$. Lower values decrease the clone population while higher values increase it.

Affinity maturation factor (ρ) — A scaling factor for the affinity maturation. Since the affinity maturation is inversely proportional to the affinity of antibodies, so is the scaling factor. Higher values decrease mutation diversity while smaller values increase mutation diversity.

2 Economic Dispatch of Electrical Energy

The economic dispatch problem (EDP) is an important power systems optimization problem with the goal of attaining optimum power dispatch for generators while respecting certain restrictions [31]. Conventional methods for the dispatch problem use Lagrangian multipliers, which take the problem's derivatives into account, yet, the generating unit's characteristics are inherently non-linear, creating multiple local minima in the cost function [27], [26], [34]. In fact, the problem is complex and highly non-linear [9], it is highly multi-modal in the prohibited zones and the valve-point loading versions [23]. The economic dispatch problem, it's variants and different solution methods have been extensively studied with several works like [19], [11], [30], [28], [29], [1], [2], [16], [3], [12], [15], [21], [4], to name a few.

Economic dispatch is the problem of deciding the most efficient, low-cost, configuration of a power system, while operational constraints, by dispatching the available power generation to meet a certain system load. It's described as a minimization problem of the total cost of power generation in a system while satisfying constraints, power limits, on the system's generation resources. There are two basic classes of constraints in the problem: i) inequality restrictions, where each generator has it's power restricted to minimum and maximum limits; ii) equality restriction,where the total generated power must be equal to the system load, or in the version with losses, must be equal to the system load plus losses. In mathematical terms, the simple version of the problem, which ignores prohibited zones and/or valve-point loading effects, is described as:

$$\min_{Pg_i} \sum_{i=1}^{Ng} C_i(Pg_i) \tag{4}$$

$$\text{subject to: } \sum_{i=1}^{Ng} Pg_i = Pd$$

$$Min(Pg_i) \leq Pg_i \leq Max(Pg_i),$$

where Ng is the number of generators, $i \in [1, Ng]$, C_i is the cost equation ($/h) of each generator, Pg_i is the active generated power (MW) of each generator and Pd is the active power demand of the system. The cost equation C_i may assume several forms, however, it's most common in the literature in the quadratic matrix equation form:como:

$$C_i(x) = x^T \cdot Q \cdot x + c^T \cdot x + \alpha, \tag{5}$$

where x is a vector with active generated power of each generator, Q is a positive-definite symmetric matrix which defines the second order elements, c is a column vector which defines the first order elements and α a column vector which defines the constants. Being a minimization problem, methods based on derivatives of the cost equations will ignore the α term.

Being a non-linear problem, iterative methods are necessary for the solution. Common iterative methods in the literature include the lambda method [32]; gradient projection [13]; linear programming [24]; dynamic programming [25]; primal-dual interior points method [22], [14] and predictor-corrector method [33]. However, it's a characteristic of non-linear problems for the computational cost to increase expressively according to the problem dimension [8].

The rest of this work is organized as follows. Section 3 describes the economic dispatch problem as an optimization problem, detailing the objective function and search space. Section 4 contains examples and presents the results of using CLONALG to the minimization of the KKT point error. Finally, the conclusion is given in section 5.

3 Optimization Problem

Since the economic dispatch problem is essentially defined by Equation 4, the search space is easily defined as a vector of generated power of each generator, that is:

$$\textit{search space} = [Pg_1, ..., Pg_{Ng}]. \tag{6}$$

Many metaheuristics, such as CLONALG, can handle lower and upper limits in the search space. These are defined in the economic dispatch problem as the inequality constraints, thus, it's possible to take advantage of CLONALG's handling of search space limits by removing the inequality constraints from the problem, changing the problem definition from Equation 4 to:

$$\begin{aligned} \min_{Pg_i} \quad & \sum_{i=1}^{Ng} C_i(Pg_i) \\ \text{subject to:} \quad & \sum_{i=1}^{Ng} Pg_i = Pd. \end{aligned} \tag{7}$$

Removing the inequality constraints from the problem definition decreases the problem complexity, leaving only the equality constraint. A common approach for metaheuristics to the economic dispatch problem is to minimize the total generator costs while considering the power demand constraint with a high valued penalty factor, such as:

$$F_{obj} = \sum_{i=1}^{Ng} C_i(Pg_i) + penalty \cdot \left(\sum_{i=1}^{Ng} Pg_i = Pd \right). \tag{8}$$

The problem with this approach is it lacks a guarantee of a global minimum. Even if the optimization algorithm, whichever is used, stalls indefinitely at a minimum, there is no numeric guarantee the global minimum has been reached. A more clever approach, which guarantees convergence to the global minimum or at least a measurement of how far away from it, is to seek the Karush-Kuhn-Tucker (KKT) point [17], [18], $\Delta_{\times}\mathcal{L}(x^*, \lambda^*) = 0$, as it has been done in [20].

3.1 KKT Point Minimization

Recalling Equation 7 and expanding it, the following is obtained:

$$\begin{aligned} \min_{x \in \Re^n} \quad & \sum_{i=1}^{Ng} Pg_i^T \cdot Q_i \cdot Pg_i + c_i^T \cdot Pg_i \\ \text{subject to:} \quad & \sum_{i=1}^{Ng} Pg_i = Pd, \end{aligned} \tag{9}$$

then utilizing the Lagrangian method to reach the minima:

$$\mathcal{L}(Pg_i, \lambda) = \sum_{i=1}^{Ng} \left(Pg_i^T \cdot Q_i \cdot Pg_i + c_i^T \cdot Pg_i \right) + \lambda^T \left(Pd - \sum_{i=1}^{Ng} Pg_i \right). \tag{10}$$

The global minimum is the set of generated power of each generator, Pg_i, and the incremental cost, λ, which satisfies the partial derivatives of Equation 10 in the origin. The new search space is then defined by:

$$search\ space = [Pg_1, ..., Pg_{Ng}, \lambda]. \tag{11}$$

Finally, the objective function for the problem is the error to the KKT point, given a set of power generation units and an incremental cost:

$$F_{obj} = \sum_{i=1}^{Ng} \left(\left| \frac{d\mathcal{L}}{dPg_i} \right| \right) + \left| \frac{d\mathcal{L}}{d\lambda} \right|. \tag{12}$$

For a system with a single generator, the objective function can be visualized in Figure 1, where it's easy to see how the global minimum is at the end of a long, narrow, almost flat valley. Finding the minimum region, the valley, is trivial, but to converge to the global minimum is, however, very difficult.

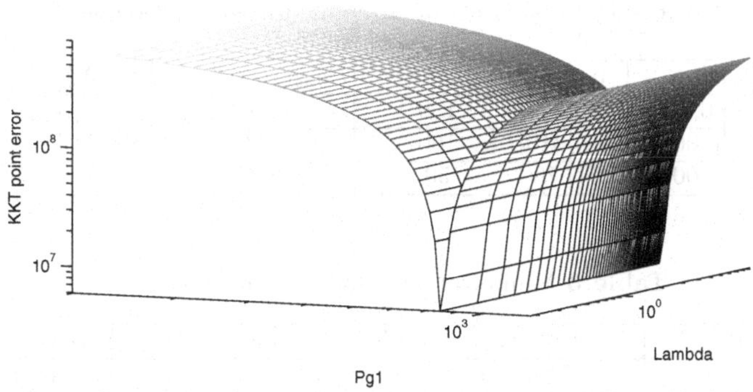

Fig. 1 Objective function for a system with a single active power generator, in logarithmic scale

4 Examples

The examples presented in this section do not take into account prohibited zones or valve-point loading effects. The maturation process of CLONALG needs to be slightly modified to increase performance in the economic dispatch problem. Considering the narrow valley of the objective function, a Gaussian distribution is more useful in the mutation process to increase local search. Also, to ensure potential solutions aren't lost, maturation only takes place in 50% of dimensions in all clones.

Results are presented for 50 runs of CLONALG for each example with the following parameters: antibody population size, $N_{abs} = 10$; selection pool size, $N_{sel} = 10$; remainder replacement pool size, $N_{new} = 0$; clonal factor, $\beta = 1$; affinity maturation factor, $\rho = 0.25$; maximum generation, $N_{gens}) = 50000$.

To assess the efficiency of CLONALG in the minimization to the KKT point error, two case studies, with and without power loss, of economic dispatch problems are analyzed.

4.1 Case Study 1

This case study, presented in Table 2, consists of three generator units with configuration given in Table 2.

The convergence results in Table 3 are presented for running CLONALG with 50000 generations, for 50 runs, showing CLONALG's capability to find the global optimum. Results for the objective function value, the error to the KKT point, and for the generation cost are also presented, showing minimum, mean, standard deviation and maximum values. Table 4 shows the best solution found by CLONALG for this case study.

Table 2 Data of case 1 of economic dispatch problem

G	a	b	c	Pg_{min} (MW)	Pg_{max} (MW)	Pd (MW)
1	0.1562	7.92	561	150	600	850
2	0.00194	7.85	310	100	400	
3	0.00482	7.97	78	50	200	

Table 3 Minimization results for the case 1

Min. F_{obj}	Mean F_{obj}	Std. Dev. of F_{obj}	Max. F_{obj}
5.4502e-08	6.7467e-06	1.3786e-05	6.6261e-05
Min. Cost ($/h)	Mean Cost ($/h)	Std. Dev. Cost ($/h)	Max. Cost ($/h)
8194.35610840853	8194.35612105488	2.0573e-006	8194.35612491201

Table 4 Best solution obtained for the case 1 in 50 runs

F^*_{obj}	Pg^*_1 (MW)	Pg^*_2 (MW)	Pg^*_3 (MW)	Incremental cost λ^* ($/MWh)
5.4502e-08	393.169843	334.603752	122.226405	9.148263

4.2 Case Study 2

This case study, presented in Table 5, is similar to Case 1, the difference lies in the power loss dependent on generator 1. The power loss changes the KKT conditions by affecting the Lagrangian derivative of the first generator and of the incremental cost, presenting a new challenge to CLONALG's convergence.

Table 5 Data of case 2 of economic dispatch problem

G	a	b	c	Pg_{min} (MW)	Pg_{max} (MW)	Pd (MW)
1	0.1562	7.92	561	150	600	$850 + 0.01359{\cdot}Pg_1$
2	0.00194	7.85	310	100	400	
3	0.00482	7.97	78	50	200	

The convergence results in Table 6 are presented for running CLONALG with 50000 generations, for 50 runs, showing CLONALG's capability to find the global optimum in a different scenario, where a power loss element based on generator 1 is introduced in the problem. Results for the objective function value, the error to the KKT point, and for the generation cost are also presented, showing minimum, mean, standard deviation and maximum values. Table 7 shows the best solution found by CLONALG for this case study.

Table 6 Minimization results for the case 2

Min. F_{obj}	Mean F_{obj}	Std. Dev. of F_{obj}	Max. F_{obj}
3.7424e-08	0.0012583	0.0085057	0.060162
Min. Cost ($/h)	Mean Cost ($/h)	Std. Dev. Cost ($/h)	Max. Cost ($/h)
8242.24097636649	8242.24667247872	0.0401672683253928	8242.52501712881

Table 7 Best solution obtained for the case 2 in 50 runs

F^*_{obj}	Pg^*_1 (MW)	Pg^*_2 (MW)	Pg^*_3 (MW)	Incremental cost λ^* ($/MWh)
3.7424e-08	374.299972	351.685238	129.101528	9.214539

5 Conclusion

This work analyzed the usage of CLONALG on the economic dispatch problem of electrical energy. The optimization problem was formulated as the minimization of the KKT point error, considering the issue that most problems in stochastic minimization have no guarantee of a global optimum. This formulation may not assure a convergence to the global optimum with stochastic optimization algorithms but, at least, provides the knowledge of how far away the solution is from the optimum and the objective function is shown to possess an easy to find minima region, although the global minimum is hard to find due to the valley created by the incremental cost. Minor modifications have been made to improve CLONALG's performance on the problem, such as changing the perturbation vector, the mutated clones, from an uniform to a Gaussian distribution with mean 0 and variance 1. In addition, a selection of dimensions similar to the binary selection used in differential evolution was used, where only 50% of the dimensions are effectivelly mutated, or maturated, reducing losses of potential solutions. The approach was tested on two case studies, where case study 1 is a simple power system with three generators and case study 2 is an extension of the first, with the addition of power losses, to test the robustness of the method. Empirically, results demonstrate the approach is successful, showing CLONALG is able to reach the global optimum on both case studies.

References

1. Abido, M.A.: A niched pareto genetic algorithm for multi objectives environmental/economic dispatch. Electrical Power and Energy Systems 25(2), 97–105 (2003)
2. Abido, M.A., Bakhashwain, J.M.: Optimal var dispatch using a multiobjective evolutionary algorithm. Electrical Power and Energy Systems 27(1), 13–20 (2005)
3. Bhatnagar, R., Rahmen, S.: Dispatch of direct load control for fuel cost minimization. IEEE Transactions on Power Systems 1(4), 96–102 (1986)

4. Zhao, B., Cao, Y.-j.: Multiple objective particle swarm optimization technique for economic load dispatch. Journal of JZhejiang University SCI 2005 6(5), 420–427 (2005)
5. Burnet, F.M.: The clonal selection theory of acquired immunity. Cambridge University Press, Cambridge (1959)
6. Castro, L.N., Zuben, F.J.V.: The clonal selection algorithm with engineering applications. In: In Proceedings of GECCO, Workshop on Artificial Immune Systems and Their Applications, Las Vegas, LA, USA, pp. 36–39 (July 2000)
7. Castro, L.N., Zuben, F.J.V.: Learning and optimization using the clonal selection principle. IEEE Transactions on Evolutionary Computation 6(3), 230–251 (2002)
8. Chandram, K., Subrahmanyam, N., Sydulu, M.: Brent method for dynamic economic dispatch with transmission losses. In: IEEE/PES Transmission and Distribution Conference and Exposition, Chicago, IL, USA (2008)
9. Coelho, L.S., Mariani, V.C.: Combining of chaotic differential evolution and quadratic programming for economic dispatch optimization with valve-point effect. IEEE Transactions on Power Systems 21(2), 989–996 (2006)
10. Cutello, V., Narzisi, G., Nicosia, G., Pavone, M.: Clonal selection algorithms: A comparative case study using effective mutation potentials. In: Jacob, C., Pilat, M.L., Bentley, P.J., Timmis, J.I. (eds.) ICARIS 2005. LNCS, vol. 3627, pp. 13–28. Springer, Heidelberg (2005)
11. Das, D.B., Patvardhan, C.: New multi-objective stochastic search technique for economic load dispatch. IEEE Proceedings on Generation, Transmission and Distribution 145(6), 747–752 (1998)
12. Dhifaoui, R., Hadj Abdallah, H., Toumi, B.: Le calcul du dispatching économique en sécurité par la méthode de continuation paramétrique. In: Séminaire à l'I.N.H, Boumerdes, Algérie (1987)
13. Granelli, G.P., Marannino, P., Montagna, M., Liew, A.C.: Fast and efficient gradient projection algorithm for dynamic generation dispatching. IEEE Proceedings on Generation, Transmission and Distribution 136(5), 295–302 (1989)
14. Granville, S.: Optimal reactive dispatch through interior point methods. IEEE Transactions on Power Systems 9(1), 136–146 (1994)
15. Guesmi, T., Hadj Abdallah, H., Ben Aribia, H., Toumi, A.: Optimisation multiobjectifs du dispatching economique / environnemental par l'approche npga. In: International Congress Renewable Energies and the Environment (CERE), Sousse, Tunisie (March 2005)
16. Hota, P.K., Dash, S.K.: Multiobjective generation dispatch through a neuro-fuzzy technique. Electric Power Components and Systems 32(11), 1191–1206 (2004)
17. Karush, W.: Minima of functions of several variables with inequalities as side constraints. Master's thesis, Department of Mathematics, University of Chicago, Chicago, Illinois, USA (2003)
18. Kuhn, H.W., Tucker, A.W.: Nonlinear programming. In: Proceedings of 2nd Berkeley Symposium, pp. 481–492. University of California Press, Berkeley (1951)
19. Lin, C.E., Viviani, G.L.: Hierarchical economic dispatch for piecewise quadratic cost functions. IEEE Transactions on Power Apparatus and Systems 103(6), 1170–1175 (1984)

20. Manoharan, P.S., Kannan, P.S., Baskar, S., Willjuice Iruthayarajan, M.: Evolutionary algorithm solution and kkt based optimality verification to multi-area economic dispatch. International Journal of Electrical Power & Energy Systems 31(7-8), 365–373 (2009)
21. Miranda, V., Hang, P.S.: Economic dispatch model with fuzzy constraints and attitudes of dispatchers. IEEE Transactions on Power Systems 20(4), 2143–2145 (2005)
22. Quintana, V.H., Torres, G.L., Medina-Palomo, J.: Interior-point methods and their applications to power systems: A classification of publications and software codes. IEEE Transactions on Power Systems 15(1), 170–176 (2000)
23. Immanuel Selvakumar, A., Thanushkodi, K.: Optimization using civilized swarm: Solution to economic dispatch with multiple minima. Electric Power Systems Research 79(1), 8–16 (2009)
24. Somuah, C.B., Khunaizi, N.: Application of linear programming re-dispatch technique to dynamic generation allocation. IEEE Transactions on Power Systems 5(1), 20–26 (1990)
25. Travers, D.L., Kaye, R.J.: Dynamic dispatch by constructive dynamic programming. IEEE Transactions on Power Systems 13(2), 72–78 (1998)
26. Victoire, T.A.A., Jeyakumar, A.E.: Reserve constraint dynamic dispatch of units with valve-point effects. IEEE Transactions on Power Systems 20(3), 1273–1282 (2005)
27. Wang, C., Shahidehpour, S.M.: Effects of ramp-rate limits on unit commitment and economic dispatch. IEEE Transactions on Power Systems 8(3), 1341–1349 (1993)
28. Wang, L., Singh, C.: Multi-objective stochastic power dispatch through a modified particle swarm optimization algorithm. In: Proceedings of IEEE Swarm Intelligence Symposium Special Session on Applications of Swarm Intelligence to Power Systems, Indianapolis, USA, pp. 127–135 (May 2006)
29. Wang, L., Singh, C.: Tradeoff between risk and cost in economic dispatch including wind power penetration using particle swarm optimization. In: International Conference on Power System Technology, Chongqing, China (2006)
30. King Warsono, D.J., Özveren, C.S.: Economic load dispatch for a power system with renewable energy using direct search method. In: 42nd International Universities Power Engineering Conference, Brighton, UK, pp. 1228–1233 (September 2007)
31. Wood, A.J., Wollenberg, B.F.: Power generation operation and control. John Wiley and Sons, New York (2004)
32. Wood, W.G.: Spinning reserve constrained static and dynamic economic dispatch. IEEE Transactions on Power Apparatus and Systems 101(2), 381–388 (1982)
33. Wu, Y., Debs, A.S., Marsten, R.E.: A direct nonlinear predictor-corrector primal-dual interior point algorithm for optimal power flow. IEEE Transactions on Power Systems 9(2), 876–883 (1994)
34. Yare, Y., Venayagamoorthy, G.K., Saber, A.Y.: Heuristic algorithms for solving convex and nonconvex economic dispatch. In: The 15th International Conference on Intelligent System Applications to Power Systems, Curitiba, Brazil (November 2009)

Dynamic Objectives Aggregation Methods in Multi-objective Evolutionary Optimization

G. Dellino[1], M. Fedele[2], and C. Meloni[3]

[1] Dipartimento di Ingegneria dell'Informazione Università degli Studi di Siena, Via Roma, 56 - 53100 Siena, Italy
dellino@dii.unisi.it

[2] Dipartimento di Scienze Economiche, Matematiche e Statistiche, Università degli Studi di Foggia, Largo Papa Giovanni Paolo II, 1 - 71100 Foggia, Italy
m.fedele@unifg.it

[3] Dipartimento di Elettrotecnica ed Elettronica, Politecnico di Bari, Via E. Orabona, 4 - 70125 Bari, Italy
meloni@deemail.poliba.it

Several approaches for solving multi-objective optimization problems entail a form of scalarization of the objectives. This chapter proposes a study of different dynamic objectives aggregation methods in the context of evolutionary algorithms. These methods are mainly based on both weighted sum aggregations and curvature variations. Since the incorporation of chaotic rules or behaviour in population-based optimization algorithms has been shown to possibly enhance their searching ability, this study proposes to introduce and evaluate also some chaotic rules in the dynamic weights generation process. A comparison analysis is presented on the basis of a campaign of computational experiments on a set of benchmark problems from the literature.

1 Introduction

A multi-objective or vector optimization problem with $m \geq 2$ objectives or criteria can be stated as follows:

$$\min_{x \in \mathcal{X}} \mathbf{f}(x) = (f_1(x), \ldots, f_m(x))' , \tag{1}$$

where $f_j : \mathbb{R}^n \rightarrow \mathbb{R}$, for $j = 1, \ldots, m$; $\mathcal{X} \subset \mathbb{R}^n$ is the set of the *feasible decision vectors*. Generally, the feasible set is implicitly given through a set of constraints. $\mathcal{Z} := \mathbf{f}(\mathcal{X})$ is the set of all values assumed by the objective function on the feasible set; it is a subset of the *objective space* $\mathbb{R}^m$ and the vectors belonging to $\mathcal{Z}$ are called *feasible criteria vectors*.

N. Nedjah et al. (Eds.): Innovative Computing Methods, SCI 357, pp. 85–103.
springerlink.com

While a multi-objective problem involving independent design functions can be solved by simply minimizing m scalar objective functions separately, when the objectives are competitive it is very difficult to obtain a single decision vector which minimizes all criteria simultaneously.

Generally, the solution of a multi-objective optimization problem is a set of vectors, the *Pareto solutions*, that provide a trade-off among criteria. The goal of a multi-objective optimizer is to achieve a set of Pareto optimal solutions. Since every Pareto point is of potential interest, the target is to *capture* the whole Pareto front.

There are several methods for solving multi-objective optimization problems; for instance, they can be solved through Evolutionary Algorithms (EAs), whose main advantage is that they can find multiple Pareto optimal solutions in a single run [2, 18, 4]. Among evolutionary methods to tackle multi-objective optimization a classical approach entails a form of scalarization of the criteria vector. Repeated applications of these methods are performed to achieve an estimation of the Pareto front. The aggregate objective function methods transform a multi-criteria optimization problem into a scalar problem using free parameters to be set; for every set of parameter values, the scalar optimization problem is solved to seek a Pareto solution. Hence, the original problem (1) is transformed as follows:

$$\min_{x \in \mathcal{X}} G(f_1(x), \ldots, f_m(x)) \,, \tag{2}$$

with $G : \mathcal{Z} \subset \mathbb{R}^m \rightarrow \mathbb{R}$.

Konak et al. in [18] provide a tutorial on adopting genetic algorithms to solve multi-objective problems, discussing also aggregation approaches based on weighted sum of the objective functions. Despite such a method is a quite common formulation for a multi-objective optimization problem, there are several issues which deserve further investigation; a recent work of Marler and Arora [23] proposes further insight on the weighted aggregation method, focusing on the choice and significance of the weights according to different criteria. Besides, [8] investigates the sensitivity of this class of methods to the weights chosen in the scalarization.

The fundamental issue of a scalarization approach is determining whether the transformed problem and the original one are equivalent. In order to provide the decision maker with the chance to choose among all optimal points, an aggregate function should be able to capture any existing Pareto solution. It is possible to prove that any Pareto optimal point can be *captured* if there is an appropriate aggregate function [24], where a point x is called *capturable* if it is a local optimum of the scalarized problem. Therefore, a main issue of this approach is the determination of an appropriate structure for the aggregate function, able to cover the Pareto front (as much as possible) by properly setting/varying the parameter values. A possible way to deal with this aspect is based on the dynamic variation of the aggregation function, through properly modifying its parameters during the optimization process.

This allows to capture multiple Pareto solutions, reducing at the same time the sensitivity issues related to a fixed choice of the weights [8].

In this work, an experimental comparative study of different Dynamic Objectives Aggregation Methods (DOAMs) in the context of evolutionary optimization algorithms is proposed; in particular, we investigate the performance of chaotic variations of the parameters (weights), based on promising results from recent research works [6, 28, 3, 5] on the integration of chaotic maps in optimization algorithms. The study is conducted on a set of benchmark problems from the literature. Section 2 presents methods based on both weighted sum aggregations, and curvature variations. In Sect. 3 the experimental setting is described. In Sect. 4, the analysis of results is reported, and some conclusions are drawn.

2 Dynamic Objectives Aggregation Methods

The aim of this section is the introduction of the evolutionary dynamic objectives aggregation methods to solve multi-objective optimization problems. Since an aggregate function maps the feasible criteria region into a one-dimensional value space, the aggregation method transforms a vector optimization problem into a scalar problem.

The most common and widely used aggregate function is the weighted sum of objectives. Although it has been shown that the weighted sum aggregate function is unable to deal with multi-objective optimization problems with a concave Pareto front, in [14, 15] it is investigated the possibility to capture also the points on concave Pareto front by using a dynamic weighted aggregation combined with evolution strategies.

In [16] the phenomenon of *global convexity* is introduced in order to explain the potential success of dynamic weighted aggregation. However, no analytical characterization is given to identify a global convex problem, therefore the discussion is based on an observed behaviour rather than theoretical analysis.

From an implementation point of view, classical methods that scalarize multiple objectives perform repeated runs in order to achieve a set of non-dominated solutions. Dynamic weighted aggregation, instead, provides an entire front of non-dominated solutions in a single run. At this aim, these procedures generally use an archive to store the non-dominated solutions obtained during the search process.

Empirical results in the literature show that the evolutionary dynamic weighted sum is able to provide a set of non-dominated solutions in one run of the evolutionary algorithm also capturing in some cases the points on concave parts of the Pareto frontier [14, 15, 16]. Besides that, a method based on the increase of the aggregate function curvature, obtained varying the exponents of the objective function in the aggregation, seems to be able to capture the points on concave regions of the front where the plain weighted sum fails.

The rationale behind an integration of the two methods can be summarized observing that by dinamically varying the curvature it may be possible to reach the concave part of the front and by dynamically varying the weights it may be possible to move close to the concave Pareto frontier.

The remainder of the section is devoted to introduce the algorithmic approaches investigated in our computational study: i) dynamic weighted sum methods, and ii) dynamic curvature variations methods.

2.1 Dynamic Weighted Sum Methods

The most widely used aggregate function is the weighted sum; the corresponding aggregate optimization problem can be stated as:

$$\min_x \sum_{i=1}^{m} w_i f_i(x) , \tag{3}$$

where w_i $(i = 1, \ldots, m)$ are non-negative weights for the corresponding objective functions f_i and $\sum_{i=1}^{m} w_i = 1$.

For every choice of the weights vector $\mathbf{w}$, the problem (3) yields an optimal Pareto point. It is well-known that a weakness of this aggregate function is the failure to capture the points on a concave Pareto fronts. In fact, it is possible to prove that every point captured by $\sum_{i=1}^{m} w_i f_i$ is in a convex region of the non-dominated frontier.

In [14] the dynamic weighted aggregation method combined with evolution strategies has been studied and it has been shown that this method is able to capture the entire Pareto frontier reaching the points in concave regions as well. This procedure is based on the dynamic aggregation approach.

While conventionally the scalarization function weights are fixed during optimization, the main idea on which the method is based is that the weights systematically change during evolution; so the function to be minimized dynamically changes. In this way the optimizer moves close to the frontier, once it achieves a non-dominated solution. During the evolution of the algorithm, the population can intersect the Pareto set, and therefore an archive is needed to record all the Pareto solutions encountered. Although it has been extensively shown that the conventional weighted sum is unable to provide the Pareto solutions on concave regions, the dynamic weighted sum method succeeds in obtaining non-dominated solutions in concave regions as well, traversing the frontier dynamically. Empirical results highlight the important role of law of the varying weights [14, 15].

Several ways of changing weights have been proposed: randomly, switching between 0 and 1, and periodically. In the first case, the weights are generated from a uniform random distribution changing in each generation. The second way of varying the weights is realized by switching them from zero to one abruptly and viceversa. Literature results suggest that the weights should vary gradually and periodically. In particular, a gradual and continuous change is needed to keep the points on a convex front: an abrupt switch of the search direction does not allow the optimizer to move close to the front, storing non-dominated points.

Since the incorporation of chaos in population-based optimization algorithms has been shown to possibly enhance their searching ability [5, 13, 22], this study proposes to introduce and evaluate also the chaotic rules in the dynamic weights generation as well.

2.2 Dynamic Curvature Variation Methods

In order to overcome the drawbacks of the weighted sum scalarization function, several aggregate functions have been introduced in the literature. In particular, to enhance the capability of objective functions to capture also the points on a concave Pareto front, in [24] it is suggested to increase the curvature of the aggregate function. A way of varying the curvature can be easily obtained applying the so called t-th power transformation to the objective function. The corresponding scalar optimization problem can be stated as follows:

$$\min_{x} \sum_{i=1}^{m} w_i (f_i(x))^t \,, \tag{4}$$

where t is a positive real number. It is found that varying all the free parameters (i.e. weights and exponents), it is possible to reach all the points on the Pareto frontier. This aggregate function is also investigated in [19, 20], where it has been proved that applying the t-th power to the objective functions the convexification of non-dominated frontier can be achieved in an appropriate equivalent objectives space. The main problem is again the choice of a function structure enabling to provide all the Pareto solutions for some values of the parameters used in aggregate function. Assuming that the aggregate objective function and the Pareto frontier satisfy certain differentiability requirements, the necessary and sufficient condition for an admissible aggregate objective function to capture the Pareto points has been developed by Messac et al. [24]. Although these conditions are inapplicable if the Pareto frontier is not known — as it is in real applications — Messac et al. suggested the use of an aggregate function (4) whose curvature can be increased by setting free parameters with the aim to enhance the capability of objective functions to capture also the points on concave Pareto front. This t-th power approach is also investigated in [19]. For sufficiently large values of t, the efficient frontier in the $[f_1^t, ..., f_m^t]$ space is guaranteed to be convex under certain conditions. Therefore, the weighted sum of the t-th power of the objectives is able to solve the problem in the $[f_1^t, ..., f_m^t]$ space. In [11] the properties of the weighted t-th power approach are summarized:

i) the optimal solutions of the t-th power problem (4) are efficient solutions of the multi-objective problem (1);
ii) for every efficient solution of the problem (1) there exists a $\widehat{t} > 0$ such that for all $t \geq \widehat{t}$ the t-th power aggregate function in (4) captures that solution.

This result guarantees the existence of a t-th power aggregate function that is able to capture the whole Pareto front. Therefore this is an important theoretical support for our work in which different rules to dynamically change the values of t are considered in addition to those concerning the weights w_i.

3 Computational Experiments

In this section the experimental setting is illustrated. The evolutionary algorithms involved in the test and their configurations are described in Subsect. 3.1. Subsection 3.2 reports on the different DOAMs considered in the experiments. In order to evaluate and compare the effectiveness of the proposed methods, a suite of test problems is employed as will be described in Subsect. 3.3.

3.1 Evolutionary Algorithms and Their Configurations

To test the methods proposed in this work, the standard genetic algorithm included in the Matlab's Genetic Algorithm and Direct Search Toolbox [29] has been used. This algorithm enables to solve single-objective optimization problems and can be easily adapted to work with dynamic objectives aggregation. Some parameters values need to be specified, before the algorithm execution [29]. In our experiments, we used default (and commonly used) values; in particular, we adopted a stochastic uniform selection operator, a scattered crossover function with probability 0.7 and a Gaussian mutation function with probability 0.3; the number of best individuals that will survive to the next population has been fixed to 2 and the stopping criterion is based on the maximum number of generations to be produced. A more refined tuning will be needed when applying these methods to specific practical applications or for a detailed study on the sensitivity of the optimization algorithm to these parameters [1].

In this work, several alternative settings have been considered for the genetic algorithm, by varying the population size (in the set $\{25, 50, 100\}$) and the number of performed iterations (100, 500 or 1000); thus, 9 overall different configurations of the genetic algorithm are used.

The considered DOAMs require to dynamically solve single-objective optimization problems, given by the dynamic weighted sum of the objectives we are really interested in. In order to keep all the non-dominated solutions obtained during the optimization process, an archive is needed. At each step a dominance analysis on the offspring population is conducted, with respect to the functions composing the aggregate objective, thus obtaining a set of individuals that are candidated to enter the archive.

The archive is composed of non-dominated solutions collected as the result of the optimization search, that has taken account not only of the aggregate function but also of its single components. It is important to update this archive, removing all dominated solutions. The update frequency of the archive (denoted by f_{upd}) is set by the user before the algorithm execution.

The archive structure A has a maximum size S, but at any iteration it may contain a number $s = |A| \leq S$ of elements. The use of the archive requires a domination analysis, a capacity control and a crowding analysis for the elements that are proposed to be enclosed in it. At each iteration the evolutionary optimizer proposes to the archive the non-dominated elements contained in its current population; they enter the archive according to a specific acceptance criterion. More specifically, the following operations are performed: a domination analysis ensures that the candidate elements are entitled to enter the archive. For each candidate two situations are possible: (a) it dominates some of the elements in A, so it replaces the dominated solutions; (b) it is non-dominated with respect to all the solutions in A, so it could be added to the archive. However, this requires a preliminary capacity control, in order to guarantee that the maximum size S is not exceeded. In case the archive size reaches the maximum value S, a crowding analysis is performed, estimating the density of solutions in the neighborhood of each element of the archive. In particular, a crowding distance value is assigned to the individuals belonging to the archive, calculated as follows: for each objective function, the solutions in the archive are sorted in ascending order of magnitude. Then an infinite distance value is assigned to solutions having the smallest and highest objective function values; for solutions other than the boundary ones, the distance value is given by

$$d_k = d_k + \frac{f_i(x_{k+1}) - f_i(x_{k-1})}{f_i^{max} - f_i^{min}} \,, \tag{5}$$

where d_k is the crowding distance associated to the k-th element ($k = 1, \ldots, |A|$), whose initial value is set to 0, $f_i(\cdot)$ is the value of the i-th objective function of the element specified in brackets, x_{k+1} and x_{k-1} are the successor and the predecessor of the k-th individual. According to the operated sorting, f_i^{max} and f_i^{min} are the maximum and minimum value of the i-th objective function. The calculation of d_k is performed with all the objective functions, finally obtaining the overall crowding distance for each individual. Thus solutions located in lesser crowded regions are preferred.

Figure 1 outlines how the algorithm works; Figure 2 illustrates the acceptance rule of new elements in the archive.

Optimization Process
Input: Parameters set by the user;

Initialization:
Generate a random initial population
begin
while the stopping criteria is not satisfied
Evaluate all individuals in the population
Rank the population, according to their fitness values
for each non-dominated individual e in the population
Archive acceptance rule(e)
if generation `mod` $f_{upd} = 0$ **then**
Update the archive, removing dominated elements
Select parents, based on their fitness
Create a new population, applying elitism, crossover and mutation operators
Update the archive, removing dominated elements
end

Fig. 1 Genetic Algorithm Scheme.

Archive acceptance rule
Input: archive A, archive size S, candidate element e;

begin
if e dominates any element in A **then**
delete all dominated elements of A and include e in A
else if no element of A dominates e and $|A| < S$ **then**
include e in A
else if no element of A dominates e and $|A| = S$ **then**
if the crowding distance of e is better than that of $x \in A$ **then**
delete x and include e in A
end

Fig. 2 The acceptance criterion of new elements in the archive.

A well-known multi-objective genetic algorithm (MOGA), NSGA-II – extensively described by Deb et al. in [10] – has also been used, aiming to compare the solutions obtained with the proposed method with those provided by a *native* MOGA. The algorithm configurations described before have also been applied to NSGA-II, except for the dominance management which is implicitly guaranteed by the algorithm itself; NSGA-II also includes a crowding analysis similar to the one we adopt (the details of NSGA-II can be found in [10]).

3.2 The Set of DOAMs

Several DOAMs have been used in the campaign of experiments conducted in this work, involving both the variation of the weights only, as in (3), and the combined variation of the weights and the exponents in (4). For the sake of simplicity, in order to illustrate the methods, a bi-objective problem is considered. In this case, the aggregate function corresponding to the k-th generation can be stated as follows:

$$G(x, k) = w_1(k) f_1^t(x) + w_2(k) f_2^t(x) \ , \tag{6}$$

where the expressions of w_1 and w_2 and the value of t depend on the adopted variation law; clearly, for $t = 1$ the simpler weighted sum aggregated function is obtained. The weights w_i can be dynamically modified according to a rule $R(k)$ described by a specific function of k:

$$w_1(k) = R(k), \qquad w_2(k) = 1 - w_1(k) \ . \tag{7}$$

The weighted aggregation methods can easily be extended to three-objective problems [17]; the weights can be generated in the following way:

$$w_1(k) = R(k) \ , \quad w_2(k) = (1 - w_1(k))R(k) \ , \quad w_3(k) = 1 - w_1(k) - w_2(k) \ .$$

We consider different rule functions, namely: *one*, *switch*, *sin*, *triangle*, *rand*, *chaos*.

The first refers to the case of fixed unitary weights (i.e. the aggregate function is simply given by the sum of the objectives). In the second case w_1 periodically changes from 0 to 1, with a given period $T = 100$ (in terms of the number of algorithm's generations). Similarly, a periodical changing of the weights can be obtained also according to a sin or triangle wave in the successive adopted rules; the sinusoidal rule is the following:

$$R(k) = |\sin(2\pi k/F)| \ , \tag{8}$$

where F is the frequency in terms of algorithm's generations. In our computational experiments, this parameter has been set to $F = 200$ following [14] showing that such a value leads to algorithms with better convergence properties. The *rand* rule, at each iteration k, generates a random value in $(0, 1)$ for w_1. The last rule applies a chaotic variation law to the weights. A logistic equation – which is extensively used to describe a chaotic system [3, 5, 21] – is employed as follows:

$$w_1(k + 1) = \mu w_1(k)(1 - w_1(k)), \qquad w_2(k) = 1 - w_1(k) \ ; \tag{9}$$

where $\mu = 4$ and $w_1(0) = 0.2027$. The previous deterministic equation (choosing $w_1(0) \in (0, 1) \backslash \{0, 0.25, 0.5, 0.75, 1\}$) shows chaotic behaviour, exhibiting a sensitive dependence on initial conditions, which is the basic characteristic of chaos. A little difference in the initial value of the chaotic variable would result in a considerable difference in its long-time behaviour. In general, this chaotic variable has special characteristics, such as ergodicity, pseudo-randomness and irregularity [3, 5, 21]. Clearly, some other well-known chaotic maps could also be employed instead of the logistic one to generate the weights in the aggregate objective functions [5].

In order to let the curvature of the aggregate function vary during the evolution process four possible strategies are proposed for the variation of the exponent. In all the cases considered in the following, the exponent value ranges between $t = 1$ and $t = 4$, retaining that greater values of t would not provide further improvements in the optimization results achieved so far.

A first scheme (*one*) considers only fixed unitary exponents. The second scheme (*step*) establishes to increment the exponent value every $N/4$ iterations, N being the maximum number of generations that can be produced. An adaptive scheme (*adapt*) has also been tested, according to which the exponent value is incremented when there is no improvement in the optimization process for a given number of iterations, which has been fixed to $\Delta = 0.05\, N$.

According to both these strategies, the exponent value is always a positive integer number. The last strategy (*cont*) considered in this work let the exponent range among the (positive) real numbers; i.e., the interval $(0, N)$ has been mapped into the interval $(1, 4)$ such that the exponent t can vary continuously in this range.

Combining the aforementioned weights-exponents strategies, 24 different algorithms are obtained. Hereinafter, each of them will be denoted indicating the strategies as an ordered pair (e.g. *chaos-step* represents the strategy with the chaotic rule for the weights and the *step* rule for exponents, respectively) and they will be labeled as reported in Table 1.

Table 1 The considered DOAMs and the corresponding weights-exponents strategies

Algorithm	Weights-Exponents Rule	Algorithm	Weights-Exponents Rule
DOAM1	chaos-one	DOAM13	chaos-adapt
DOAM2	one-one	DOAM14	one-adapt
DOAM3	rand-one	DOAM15	rand-adapt
DOAM4	switch-one	DOAM16	switch-adapt
DOAM5	sin-one	DOAM17	sin-adapt
DOAM6	triangle-one	DOAM18	triangle-adapt
DOAM7	chaos-step	DOAM19	chaos-cont
DOAM8	one-step	DOAM20	one-cont
DOAM9	rand-step	DOAM21	rand-cont
DOAM10	switch-step	DOAM22	switch-cont
DOAM11	sin-step	DOAM23	sin-cont
DOAM12	triangle-step	DOAM24	triangle-cont

3.3 Test Problems

The computational test of the methods has been conducted on a set of benchmark problems, characterized by different specific features in the Pareto front, so that the general results obtained would not depend on the particular test problem chosen. Problems P_1-P_7 are discussed by Jin et al. in [14, 15, 16]: in P_2-P_5 it is assumed that $x_i \in [0,1]$ for all $i = 1, \ldots, n$; while in P_1, P_6, and P_7 there are no restrictions on the range of the decision variables. The problem P_1 has the following objective functions:

$$f_1 = \frac{1}{n}\sum_{i=1}^{n} x_i^2 \,, \qquad f_2 = \frac{1}{n}\sum_{i=1}^{n}(x_i - 2)^2 \,; \tag{10}$$

and produces a uniform Pareto front. P_2 is described by:

$$\begin{aligned} f_1 &= x_1 \\ g(x_2, \ldots, x_n) &= 1 + \frac{9}{n-1}\sum_{i=2}^{n} x_i \,, \\ f_2 &= g \times (1 - \sqrt{f_1/g}) \,. \end{aligned} \tag{11}$$

having a convex Pareto front. Because of the interest in studying problems showing non-convex or discontinuous Pareto front, some instances belonging to this class have been considered. The following problem, P_3, has a concave Pareto front and is defined as follows:

$$\begin{aligned} f_1 &= x_1 \,, \\ g(x_2, \ldots, x_n) &= 1 + \frac{9}{n-1}\sum_{i=2}^{n} x_i \,, \\ f_2 &= g \times \left(1 - (f_1/g)^2\right) \,. \end{aligned} \tag{12}$$

The fourth problem, P_4, has been obtained through combining — in some sense — P_2 and P_3:

$$\begin{aligned} f_1 &= x_1 \,, \\ g(x_2, \ldots, x_n) &= 1 + \frac{9}{n-1}\sum_{i=2}^{n} x_i \,, \\ f_2 &= g \times \left(1 - \sqrt[4]{f_1/g} - (f_1/g)^4\right) \,. \end{aligned} \tag{13}$$

Its Pareto front is neither purely convex nor purely concave. The following problem, P_5, is characterized by a Pareto front consisting of a number of separated convex parts.

$$
\begin{aligned}
f_1 &= x_1 , \\
g(x_2, \ldots, x_n) &= 1 + \frac{9}{n-1} \sum_{i=2}^{n} x_i , \\
f_2 &= g \times \left(1 - \sqrt{f_1/g} - (f_1/g)\sin(10\pi f_1)\right) .
\end{aligned}
\tag{14}
$$

Problem P_6 is defined through the following equations:

$$
\begin{aligned}
f_1 &= 1 - \exp\left\{ - \sum_{i=1}^{n} \left(x_i - \frac{1}{\sqrt{n}} \right)^2 \right\} \\
f_2 &= 1 - \exp\left\{ - \sum_{i=1}^{n} \left(x_i + \frac{1}{\sqrt{n}} \right)^2 \right\} ,
\end{aligned}
\tag{15}
$$

showing a concave Pareto front. Another problem, P_7, is taken from [16] extending one of the test function proposed in the literature [25] to the two-dimensional case:

$$
\begin{aligned}
f_1 &= \exp(-x_1) + 1.4 \exp(-x_1^2) + \exp(-x_2) + 1.4 \exp(-x_2^2) , \\
f_2 &= \exp(x_1) + 1.4 \exp(-x_1^2) + \exp(x_2) + 1.4 \exp(-x_2^2) .
\end{aligned}
\tag{16}
$$

The resulting Pareto front is continuous and non-convex. Even if the problem is considered an easy task for evolutionary optimizers [16], the region that defines the Pareto front in the parameter space is disconnected; so, it could be an interesting problem to be studied. In the following, two other benchmark problems, P_8-P_9, are considered, because of the particular shape of their feasible region and/or Pareto front; these problems are described in [27]. Problem P_8 is referred to as the TNK problem. The objectives are very simple, and defined by

$$
f_1 = x_1, \qquad f_2 = x_2 , \tag{17}
$$

where

$$
x_1 \in [0, \pi], \qquad x_2 \in [0, \pi] .
$$

The constraints are

$$
\begin{aligned}
g_1 &= -x_1^2 - x_2^2 + 1 + 0.1 \cos(16 \arctan(x_2/x_1)) \leq 0 , \\
g_2 &= (x_1 - 0.5)^2 + (x_2 - 0.5)^2 \leq 0.5 .
\end{aligned}
\tag{18}
$$

Problem P_9 is the so-called Poloni's test problem: the objective functions are defined as

$$
f_1 = 1 + (a - b)^2 + (c - d)^2 , \qquad f_2 = (x_1 + 3)^2 + (x_2 + 1)^2 , \tag{19}
$$

where the parameters introduced in the expression of f_1 are:

$$\begin{aligned} a &= 0.5\sin(1) - 2.0\cos(1) + 1.0\sin(2) - 1.5\cos(2)\ , \\ b &= 0.5\sin(x_1) - 2.0\cos(x_1) + 1.0\sin(x_2) - 1.5\cos(x_2)\ , \\ c &= 1.5\sin(1) - 1.0\cos(1) + 2.0\sin(2) - 0.5\cos(2)\ , \\ d &= 1.5\sin(x_1) - 1.0\cos(x_1) + 2.0\sin(x_2) - 0.5\cos(x_2)\ . \end{aligned}$$

The variables ranges are: $x_1 \in [-\pi, \pi]$ and $x_2 \in [-\pi, \pi]$.

4 Discussion of Results

Appropriate metrics must be selected in order to evaluate the behaviour of the considered algorithms. The literature offers different metrics to measure the performance of algorithms for multi-criteria optimization problems. Nevertheless, no single metric is able to assess the total algorithmic performance. The metrics adopted in this study are reported below. Clearly, these metrics should not be considered as a complete list of all possible metrics. For instance, in our computational experiments, we do not consider particular temporal metrics, limiting our analysis only to the computation times required by the algorithm. Although used with cases with few objectives, the considered metrics can also be applied when a larger number of objectives is considered.

We are interested in measuring how far the non-dominated solutions obtained by the algorithm, i.e. the solution front (SF), are from the ideal point (IP). An ideal point (IP) is defined as a point characterized by the best values for each objective. We also use the concept of the nadir point (NP) which is defined as a point characterized by the worst values for each objective. To this aim, the adopted measure is given by an hyperarea or hypervolume ratio. This metric requires the knowledge (i.e. the computation) of an ideal point (IP) and a nadir point (NP) for the problem. The ideal point and the nadir point can be viewed as vertices of a regular polytope defining a hypervolume (A_t), i.e. the total area, in the space of the objectives (a rectangle in the two-objective case). The *hyperarea ratio* (HR) performance measure is defined as

$$HR = A_d/A_t\ , \tag{20}$$

where A_d indicates the dominated area between the nadir point NP and the solution front SF, as proposed by Fleischer [12]. A large value of HR is expected from a good algorithm. The hyperarea ratio is a good performance indicator, but it does not take into account how the efficient points are distributed on the estimated solution front.

Therefore as secondary indicators, we report the number of non-dominated elements (ND) and the *spacing* (S). The last one is a metric measuring the spread (distribution) of vectors throughout the solution front (SF). We refer to the definition reported in [31] for two objective functions: the spacing measures the range variance of neighboring vectors in SF:

$$S = \sqrt{\frac{1}{N-1}\sum_{i=1}^{N}(\overline{d} - d_i)^2}\ , \tag{21}$$

where $d_i = \min_j(|f_1^i(x) - f_1^j(x)| + |f_2^i(x) - f_2^j(x)|)$, $i, j = 1, \ldots, N$; $\overline{d}$ is the mean of all d_i, and N is the number of vectors in SF. A value of zero for this metric indicates that all non-dominated solutions are equally spaced. It is worth observing that no single metric can completely capture total multi-objective evolutionary algorithm performance [31].

For each experiment, three different runs have been executed initializing algorithms with random populations. Tables 2 and 3 contain the average results on 3 runs for each of the 9 algorithms configurations described in Subsect. 3.1, so each entry is the average over 27 experimental values.

Table 2 Average values of HR (in %), ND, and S achieved by the algorithms for problems P_1-P_5

Algorithms	P_1			P_2			P_3			P_4			P_5		
	HR	ND	S	HR	ND	S	HR	ND	S	HR	ND	S	HR	ND	S
DOAM1	74	25.37	0.36	90	10.67	0.42	85	6.26	0.17	81	7.00	0.29	86	6.04	0.21
DOAM2	74	24.85	0.38	91	7.59	0.45	83	4.33	0.29	81	7.04	0.25	83	5.19	0.40
DOAM3	75	25.41	0.29	91	8.52	0.32	84	4.96	0.35	81	6.33	0.29	84	5.26	0.56
DOAM4	73	23.52	0.53	88	3.48	0.18	82	2.22	0.13	81	5.78	0.29	83	3.78	0.42
DOAM5	75	24.93	0.31	91	8.07	0.24	85	4.44	0.18	81	6.37	0.25	86	5.07	0.33
DOAM6	76	24.15	0.33	90	6.11	0.37	85	5.22	0.18	81	6.74	0.27	85	5.74	0.36
DOAM7	75	25.07	0.36	90	8.04	0.30	84	5.11	0.32	81	6.41	0.25	85	6.44	0.36
DOAM8	75	25.00	0.38	90	7.78	0.28	84	4.04	0.30	81	7.48	0.31	83	5.41	0.55
DOAM9	76	24.70	0.31	91	6.93	0.40	84	4.33	0.29	81	6.67	0.27	85	5.22	0.36
DOAM10	75	23.52	0.46	89	3.67	0.16	82	2.59	0.23	80	5.48	0.35	84	3.85	0.34
DOAM11	75	24.63	0.32	91	7.22	0.28	84	4.11	0.17	81	7.15	0.29	85	5.44	0.27
DOAM12	75	25.44	0.28	91	8.33	0.31	84	4.67	0.31	81	7.52	0.27	85	5.33	0.32
DOAM13	74	24.85	0.32	91	8.96	0.20	84	4.89	0.21	81	7.15	0.26	85	5.78	0.28
DOAM14	76	25.93	0.32	91	7.41	0.25	84	3.89	0.25	81	7.44	0.28	83	5.33	0.60
DOAM15	73	25.33	0.31	90	6.07	0.30	85	4.70	0.14	81	6.78	0.31	83	6.15	0.42
DOAM16	73	24.26	0.59	88	3.04	0.13	83	3.00	0.18	80	4.74	0.37	84	4.07	0.32
DOAM17	75	25.30	0.27	91	6.67	0.13	85	4.85	0.14	81	6.74	0.32	86	5.48	0.41
DOAM18	76	25.15	0.38	91	8.52	0.25	84	4.56	0.32	82	7.44	0.23	85	6.26	0.31
DOAM19	75	24.93	0.30	90	8.52	0.25	84	5.22	0.41	81	7.11	0.29	85	6.63	0.27
DOAM20	75	25.22	0.33	91	7.93	0.23	84	5.07	0.37	81	7.11	0.31	85	5.74	0.40
DOAM21	75	23.89	0.38	91	7.07	0.31	83	4.19	0.31	81	6.78	0.30	85	6.11	0.43
DOAM22	74	23.85	0.48	89	3.74	0.26	82	3.04	0.10	80	5.48	0.32	83	3.96	0.12
DOAM23	75	24.81	0.33	91	7.63	0.36	84	5.48	0.58	81	7.15	0.25	85	5.89	0.32
DOAM24	76	25.59	0.33	90	7.30	0.42	84	5.85	0.30	81	7.19	0.22	85	5.89	0.49
NSGA-II	65	26.52	0.07	90	4.19	0.17	84	2.59	0.15	81	4.96	0.21	86	4.11	0.25

The collected results show that methods based on periodical changes of weights often achieve relatively good performances with respect to other DOAMs as well as to a state-of-the-art (native) multi-objective optimizer. This behaviour seems to be reinforced by the use of exponents variation. Nonetheless, it is noticeable the competitive performance of the DOAMs based on a *chaos* rule in terms of HR, ND, and S. On the other hand, strategies based on the *switch* rule (no matter what strategy is adopted for exponents) almost always give relatively modest results.

Although each average is composed of a large number of data points, it is necessary to carry out a statistical analysis to assess if the observed differences in the average values are indeed statistically significant. Even if in the optimization literature, parametric test methods are not often adopted due to the assumptions that the data have to satisfy and nonparametric test methods are in general preferred because they are distribution-free, the former ones result more powerful allowing for a much deeper analysis of data. In fact, in non-parametric testing a lot of information is lost because the data

Table 3 Average values of HR (in %), ND, and S achieved by the algorithms for problems P_6-P_9

Algorithms	P_6			P_7			P_8			P_9		
	HR	ND	S	HR	ND	S	HR	ND	S	HR	ND	S
DOAM1	21	16.79	0.14	100	328.19	3.41	23	4.70	0.18	91	224.79	0.21
DOAM2	21	15.29	0.14	100	152.93	1.5e56	21	4.89	0.12	89	126.25	0.63
DOAM3	21	16.13	0.18	100	310.37	3.68	22	4.63	0.14	91	212.92	0.27
DOAM4	22	14.96	0.14	100	474.59	1.7e11	21	4.48	0.18	82	36.63	1.56
DOAM5	22	19.17	0.11	100	371.56	8.98	22	4.67	0.16	91	160.58	1.18
DOAM6	23	20.46	0.10	100	327.00	4.82	21	4.48	0.16	91	166.29	1.29
DOAM7	21	17.46	0.16	100	329.52	5.31	21	4.11	0.14	91	244.50	0.23
DOAM8	21	15.08	0.18	100	141.93	3.54	22	5.00	0.13	91	158.46	0.27
DOAM9	21	17.04	0.12	100	315.37	3.78	21	4.37	0.16	91	231.00	0.26
DOAM10	21	15.92	0.12	100	475.74	1.8e10	22	4.59	0.15	83	38.71	1.60
DOAM11	23	20.83	0.15	100	373.44	9.37	21	4.30	0.18	91	176.75	0.68
DOAM12	24	20.21	0.11	100	325.22	3.70	22	4.15	0.12	91	184.75	0.60
DOAM13	22	17.96	0.11	100	339.44	3.90	22	4.48	0.14	91	243.42	0.41
DOAM14	20	16.75	0.14	100	144.70	4.08	22	4.78	0.13	91	142.42	0.27
DOAM15	21	16.79	0.12	100	319.78	3.92	21	4.15	0.12	91	218.13	0.24
DOAM16	21	15.63	0.14	100	478.63	1.3e12	21	4.11	0.16	85	39.46	0.96
DOAM17	24	21.21	0.08	100	375.22	11.70	20	4.30	0.14	91	171.42	0.89
DOAM18	23	19.13	0.14	100	332.15	4.19	22	4.59	0.19	91	196.75	0.91
DOAM19	21	17.08	0.13	100	287.00	4.55	21	4.15	0.12	91	210.83	0.26
DOAM20	21	15.54	0.20	100	93.52	3.27	22	4.48	0.15	91	148.50	0.29
DOAM21	22	17.17	0.14	100	256.85	4.45	23	4.56	0.15	91	184.96	0.27
DOAM22	22	16.38	0.13	100	482.89	2.9e10	21	4.30	0.11	80	32.25	1.17
DOAM23	23	19.25	0.10	100	290.78	5.85	22	4.48	0.17	91	162.50	1.08
DOAM24	23	20.17	0.11	100	258.74	3.49	22	4.33	0.13	91	171.58	1.20
NSGA-II	18	16.92	0.03	100	411.33	1.5e11	16	2.48	0.12	91	440.42	0.16

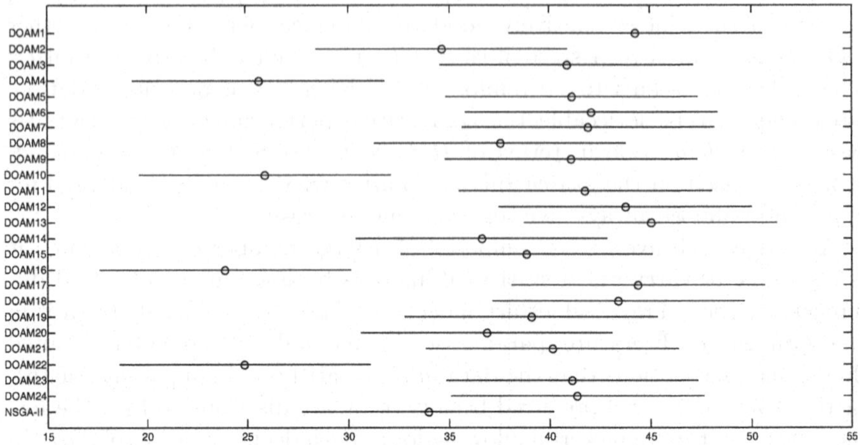

Fig. 3 Means plot and Tukey's HSD confidence intervals ($\alpha = 0.05$) resulting from the Rank-based Friedman analysis on HR

have to be ranked and the differences in the values are transformed into a rank value; therefore, besides the non-parametric rank-based Friedman's test, we consider parametric ANOVA analysis [7, 9, 26].

In these analyses we assume only HR as response variable.

Figures 3 and 4 show the means plot in the Friedman and ANOVA analysis, respectively. The analyses are conducted on the algorithm factor considering 24 different DOAMs and NSGA-II with three replicates for each experiment which is characterized by the problem and by the algorithm configuration (a total of 81 combinations are considered). Thus, in our Friedman analysis, for every experiment 75 ranks are obtained, assigning a larger rank to better results. For both tests we use a 95% confidence interval and adopt the Tukey's HSD intervals [30, 9, 26]. As it can be seen in Figures 3 and 4, the non-parametric test is less powerful neglecting the differences in the response variables and presenting much wider confidence intervals. In the reported plots, overlapping confidence intervals indicate a non-statistically significant difference on the average performance of the respective algorithms.

These analyses clearly confirm the negative assessment on *switch* strategies and the promising behaviour of chaos-based DOAMs which are often the best strategy (e.g. see DOAM13 in Figure 3), providing good performance based on the three metrics considered in Tables 2 and 3 and being non-dominated with respect to the other strategies (see Figures 3 and 4). This result seems to support the increasing research interest in the introduction of some form of chaotic behaviour in stochastic optimizers [3, 5, 13, 22].

These experimental results are of interest also in different contexts such as: the development of multi-objective optimization algorithms starting from well-established evolutionary single-objective optimizers; the design of com-

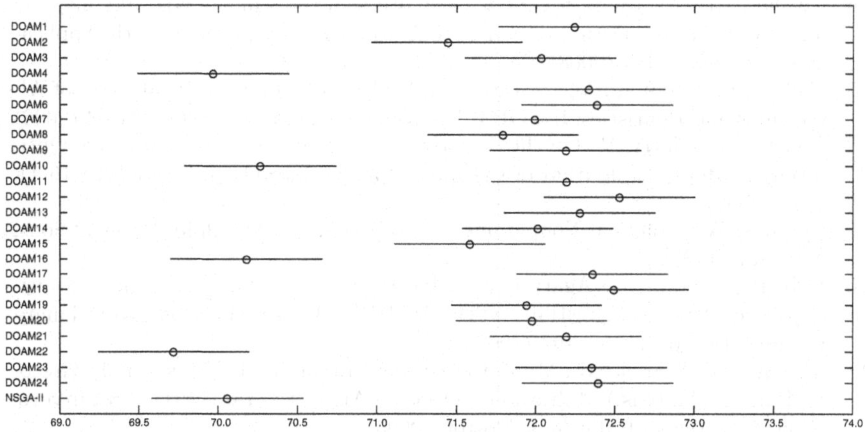

Fig. 4 Means plot and Tukey's HSD confidence intervals ($\alpha = 0.05$)resulting from the ANOVA analysis on HR

pact (and fast) local search procedures; surrogate-based optimization; landscape approximation of costly functions. Moreover, the case of bi-objective optimization could be of interest in Mean-Variance optimization problems often used to model either the risk preferences of the decision maker or robustness requirements.

The observations based on the encouraging results from the conducted experiments indicate that different aspects deserve further research efforts including the extension to the experimental campaign on a wider set of problems (also from real applications); the possible use of different quality indicators; the comparison of DOAMs with other state-of-the-art MOGAs; the investigation of the effects of the interactions of weights and exponents based rules; and the consideration of other chaotic DOAMs in order to deeply investigate their effectiveness.

References

1. Bartz-Beielstein, T.: Experimental Research in Evolutionary Computation–The New Experimentalism. Natural Computing Series. Springer, Heidelberg (2006)
2. Branke, J., Deb, K., Miettinen, K., Słowiński, R.: Multiobjective Optimization. In: Interactive and Evolutionary Approaches, Springer, Heidelberg (2008)
3. Bucolo, M., Caponetto, R., Fortuna, L., Frasca, M., Rizzo, A.: Does chaos work better than noise? IEEE Circuits and Systems Magazine 2(3), 4–19 (2002)
4. Burke, E.K., Landa Silva, J.D.: The influence of the fitness evaluation method on the performance of multiobjective search algorithms. European Journal of Operational Research 169, 875–897 (2006)
5. Caponetto, R., Fortuna, L., Fazzino, S., Xibilia, M.G.: Chaotic Sequences to Improve the Performance of Evolutionary Algorithms. IEEE Transactions On Evolutionary Computation 7(3), 289–304 (2003)

6. Coelho, L.d.S., Mariani, V.C.: Use of chaotic sequences in a biologically inspired algorithm for engineering design optimization. Expert Systems with Applications 34, 1905–1913 (2008)
7. Coffin, M., Saltzman, M.J.: Statistical Analysis of Computational Tests of Algorithms and Heuristics. INFORMS Journal on Computing 12(1), 24–44 (2000)
8. Collette, Y., Siarry, P.: On the Sensitivity of Aggregative Multiobjective Optimization Methods. Journal of Computing and Information Technology 16(1), 1–13 (2008)
9. Conover, W.: Practical Nonparametric Statistics, 3rd edn. John Wiley & Sons, New York (1999)
10. Deb, K., Pratap, A., Agarwal, S., Meyarivan, T.: A Fast and Elitist Multiobjective Genetic Algorithm: NSGA-II. IEEE Transactions on Evolutionary Computation 6(2), 182–197 (2002)
11. Ehrgott, M., Wiecek, M.: Multiobjective programming. In: Figueira, J., Greco, S., Ehrgott, M. (eds.) Multicriteria Decision Analysis: State of the Art Surveys, Kluwer Academic Publishers, Boston (2005)
12. Fleischer, M.: The Measure of Pareto Optima: Applications to Multi-objective Metaheuristics. LNCS, pp. 519–533. Springer, Berlin (2003)
13. Greenwood, G.W.: Chaotic behavior in evolution strategies. Physica D 109, 343–350 (1997)
14. Jin, Y., Olhofer, M., Sendhoff, B.: Dynamic weighted aggregation for evolutionary multi-objective optimization: why does it work and how? In: Proceedings of the Genetic and Evolutionary Computation Conference, San Francisco (2001)
15. Jin, Y., Okabe, T., Sendhoff, B.: Adapting weighted aggregation for multiobjective evolution strategies. LNCS, pp. 96–110. Springer, Zurich (1993)
16. Jin, Y.: Effectiveness of weighted aggregation of objectives for evolutionary multiobjective optimization: methods, analysis and applications (2002) (unpublished manuscript)
17. Jin, Y., Okabe, T., Sendhoff, B.: Solving Three-objective Optimization Problems Using Evolutionary Dynamic Weighted Aggregation: Results and Analysis. In: Cantú-Paz, E., Foster, J.A., Deb, K., Davis, L., Roy, R., O'Reilly, U.-M., Beyer, H.-G., Kendall, G., Wilson, S.W., Harman, M., Wegener, J., Dasgupta, D., Potter, M.A., Schultz, A., Dowsland, K.A., Jonoska, N., Miller, J., Standish, R.K. (eds.) GECCO 2003. LNCS, vol. 2723, pp. 636–638. Springer, Heidelberg (2003)
18. Konak, A., Coit, D.W., Smith, A.E.: Multi-objective optimization using genetic algorithms: A tutorial. Reliability Engineering and System Safety 91, 992–1007 (2006)
19. Li, D.: Convexification of a noninferior frontier. Journal of Optimization Theory and Applications 88(1), 177–196 (1996)
20. Li, D., Biswal, M.P.: Exponential transformation in convexifying a noninferior frontier and exponential generating method. Journal of Optimization Theory and Applications 99(1), 183–199 (1998)
21. Liu, B., Wang, L., Jin, Y.-H., Tang, F., Huang, D.-X.: Improved particle swarm optimization combined with chaos. Chaos, Solitons and Fractals 25, 1261–1271 (2005)
22. Lu, H., Zhang, H., Ma, L.: A new optimization algorithm based on chaos. Journal of Zhejiang University SCIENCE A 7(4), 539–542 (2006)
23. Marler, R.T., Arora, J.S.: The weighted sum method for multi-objective optimization: new insights. Structural Multidisciplinary Optimization 41, 853–862 (2010)

24. Messac, A., Sukam, C.P., Melachrinoudis, E.: Aggregate objective functions and Pareto frontiers: required relationships and practical implications. Optimization and Engineering 1, 171–188 (2000)
25. Messac, A., Sundararaj, G.J., Tappeta, R.V., Renaud, J.E.: Ability of objective functions to generate points on nonconvex pareto frontiers. AIAA Journal 38(6), 1084–1091 (2000)
26. Montgomery, D.C.: Design and Analysis of Experiments, 6th edn. John Wiley & Sons, New York (2004)
27. Rigoni, E., Poles, S.: NBI and MOGA-II, two complementary algorithms for Multi-Objective optimizations. In: Branke, J., Deb, K., Miettinen, K., Steuer, R.E. (eds.) Practical Approaches to Multi-Objective Optimization. Dagstuhl Seminar Proceedings, vol. 4461 (2005)
28. Tavazoei, M.S., Haeri, M.: Comparison of different one-dimensional maps as chaotic search pattern in chaos optimization algorithms. Applied Mathematics and Computation 187, 1076–1085 (2007)
29. The MathWorks, Inc. Genetic Algorithm and Direct Search Toolbox, Natick, Massachussetts (2004)
30. The MathWorks, Inc. Statistics Toolbox, Natick, Massachussetts (2007)
31. Van Veldhuizen, A.D., Lamont, G.B.: On measuring Multiobjective Evolutionary Algorithm Performance. In: Proceedings of IEEE Congress on Evolutionary Computation - CEC, pp. 204–211 (2000)

Evolutionary IP Mapping for Efficient NoC-Based System Design Using Multi-objective Optimization

Nadia Nedjah[1], Marcus Vinícius Carvalho da Silva[1], and Luiza de Macedo Mourelle[2]

[1] Department of Electronics Engineering and Telecommunications
[2] Department of Systems Engineering and Computation, Engineering Faculty, State University of Rio de Janeiro, Brazil

Summary. Network-on-chip (NoC) are considered the next generation of communication infrastructure, which will be omnipresent in most of industry, office and personal electronic systems. In the platform-based methodology, an application is implemented by a set of collaborating intellectual properties (IPs) blocks. In this paper, we use multi-objective evolutionary optimization to address the problem of mapping topologically pre-selected sets IPs, which constitute the set of optimal solutions that were found for the IP assignment problem, on the tiles of a mesh-based NoC. The IP mapping optimization is driven by the area occupied, execution time and power consumption.

1 Introduction

As the integration rate of semiconductors increases, more complex cores for *system-on-chip* (SoC) are launched. A simple SoC is formed by homogeneous or heterogeneous independent components while a complex SoC is formed by interconnected heterogeneous components. The interconnection and communication of these components form a *network-on-chip* (NoC). A NoC is similar to a general network but with limited resources, area and power. Each component of a NoC is designed as an *intellectual property* (IP) block. An IP block can be of general or special purpose such as processors, memories and DSPs [7].

Normally, a NoC is designed to run a specific application. This application, usually, consists of a limited number of tasks that are implemented by a set of IP blocks. Different applications may have a similar, or even the same, set of tasks. An IP block can implement more than a single task of the application. For instance, a processor IP block can execute many tasks as a general processor does but a multiplier IP block for floating point numbers can only multiply floating point numbers. The number of IP blocks designers, as well as the number of available IP blocks, is growing up fast.

N. Nedjah et al. (Eds.): Innovative Computing Methods, SCI 357, pp. 105–129.
springerlink.com

In order to yield an efficient NoC-based design for a given application, it is necessary to choose the adequate minimal set of IP blocks. With the increase of IP blocks available, this task is becoming harder and harder. Besides IP blocks carefully assignment, it is also necessary to map the blocks onto the NoC available infra-structure, which consists of a set of *cores* communicating through *switches*. A bad mapping can degrade the NoC performance. Different optimization criteria can be pursued depending on how much information details is available about the application and IP blocks.

Usually, the application is viewed as a graph of tasks called *task graph* (TG). The IP blocks features can be obtained from their companion documentation. The IP assignment and IP mapping are key research problems for efficient NoC-based designs [15]. These two problems can be solved using multi-objective optimizations in which some of the objectives are conflicting. Because of their nature, both IP assignment and mapping are classified as *NP*-hard problems [6]. Normally, deterministic techniques are not viable to solve such problems, so we use multi-objective evolutionary algorithms (MOEAs) with specific operators and objective functions to yield an optimal IP mapping for a prescribed set of IP assignments. These constitute the set of optimal solutions that were found in the IP assignment satge. For this purpose, one needs to select the best minimal set of objectives to be optimized. The wrong set of optimized objectives can lead to average instead of best results. Here, we assume that the IP assignment has been performed and is available for mapping the application TG onto the NoC infrastructure.

In this paper, we propose a multi-objective evolutionary-based decision support system to help NoC designers. For this purpose, we propose a structured representation of the TG and an IP repository that will feed data into the system. We use the data available in the Embedded Systems Synthesis benchmarks Suite (E3S) [3] as our IP repository. The E3S is a collection of TGs, representing real applications based on embedded processors from the Embedded Microprocessor Benchmark Consortium (EEMBC). It was developed to be used in system-level allocation, assignment, and scheduling research. We used two MOEAs: NSGA-II [2] and microGA [1]. Both of these algorithms were modified according to some prescribed NoC design constraints.

The rest of the paper is organized as follows: First, in Section 2, we present briefly some related research work. Then, in Section 3, we introduce an overview of NoC structure. Subsequently, in Section 4, we describe a structured TG and IP repository model based on the E3S data. After that, in Section 5, we introduce the mapping problem in NoC-based environments. Then, in Section 7, we sketch the two MOEAs used in this work, individual representations and objective functions for the optimization stage. Later, in Section 9, we show some experimental result. Last but not least, in Section 10, we draw some conclusions and outline new directions for future work.

2 Related Work

The problems of mapping IP blocks into a NoC physical structure have been addressed in some previous studies with different emphasis. Some of these works did not take into account of the multi-objective nature of these problems and adopted a single objective optimization approach.

Hu and Marculescu [7] proposed a branch and bound algorithm which automatically maps IPs/cores into a mesh based NoC architecture that minimizes the total amount of consumed power by minimizing the total communication among the used cores. Specified constraints through bandwidth reservation were defined to control communication limits.

Lei and Kumar [11] proposed a two step genetic algorithm for mapping the TG into a mesh based NoC architecture that minimizes the execution time. In the first step, they assumed that all communication delays are the same and selected IP blocks based on the computation delay imposed by the IPs only. In the second step, they used real communication delays to find an optimal binding of each task in the TG to specific cores of the NoC.

Murali and De Micheli [13] addressed the problem under the bandwidth constraint with the aim of minimizing communication delay by exploiting the possibility of splitting traffic among various paths. Splitting the traffic increases the size of the routing component at each node but the authors were not worried about size.

Zhou et al. [21] suggested a multi-objective exploration approach, treating the mapping problem as a two conflicting objective optimization problem that attempts to minimize the average number of hops and achieve a thermal balance. The number of hops is incremented every time a data cross a switch before reaching its target. They used NSGA [18], multi-objective evolutionary algorithm. They also formulated a thermal model to avoid hot spots, which are areas with high computing activity.

Jena and Sharma [9] addressed the problem of topological mapping of IPs/cores into a mesh-based NoC in two systematic steps using the NSGA-II [2]. The main objective was to obtain a solution that minimizes the energy consumption due to both computational and communicational activities and also minimizes the link bandwidth requirement under some prescribed performance constraints.

As a recent field of research, the available literature about NoC-based design optimization is scarce. The aforementioned works represent the state of the art of this field. In [7], [11] and [13], only one objective was considered and only [11] used an evolutionary approach. In [21] and [9], two objectives were considered and both adopted a MOEA to solve the problem.

3 NoC Internal Structure

A NoC platform consisting of architecture and design methodology, which scales from a few dozens to several hundreds or even thousands of resources [10]. As mentioned before, a resource may be a processor core, DSP core, an FPGA block, a dedicated hardware block, mixed signal block, memory block of any kind such as RAM, ROM or CAM or even a combination of these blocks.

A NoC consists of set of *resources* (R) and *switches* (S). Resources and switches are connected by *channels*. The pair (R, S) forms a *tile*. The simplest way to connect the available resources and switches is arranging them as a mesh so these are able to communicate with each other by sending messages via an available path. A switch is able to buffer and route messages between resources. Each switch is connected to up to four other neighboring switches through input and output channels. While a channel is sending data another channel can buffer incoming data. Note that energy consumption is proportional to the number of message exchanges. Therefore, the communication between resources and distance between them must be considered during the mapping stage. Fig. 1 shows the architecture of a mesh-based NoC where each resource contains one or more IP blocks (RNI for resource network interface, D for DSP, M for memory, C for cache, P for processor, FP for floating-point unit and Re for reconfigurable block). Besides the mesh topology, there are more complex topologies like *torus*, *hypercube*, 3-*stage clos* and *butterfly* [14]. Note that mesh-based NoC does not always represent the best topological choice.

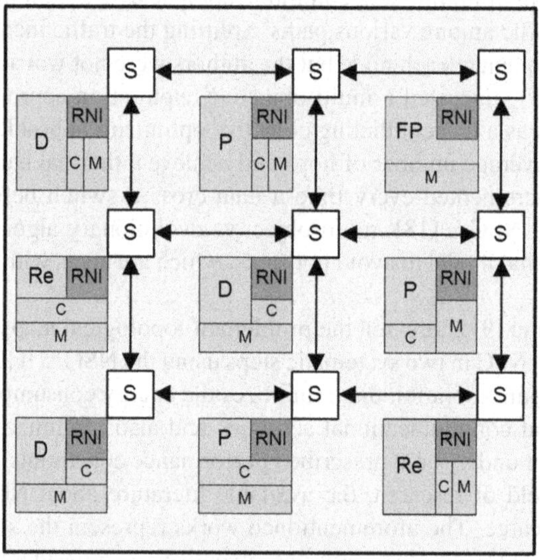

Fig. 1 Mesh-based NoC with 9 resources

Every resource has an unique identifier and is connected to the network via a switch. It communicates with the switch through the available RNI. Thus, any set of IP blocks can be plugged into the network if its footprint fits into an available resource and if this resource is equipped with an adequate RNI. The NoC implements the lower four layers from OSI model [8]: physical, data-link, network and transport layers. The RNI implements all four layers and it packs transport layer messages into network layer packets; so, the switch-to-switch interface implements only the three lower protocol layers. The basic communication mechanism envisioned among computing resources is message passing. However, it is possible to add additional protocols on top of the transport layer.

4 Task Graph and IP Repository Models

In order to formulate the IP mapping problem, it is necessary to introduce a formal definition of an application first. An application can be viewed as a set of tasks that can be executed sequentially or in parallel. It can be represented by a directed graph of tasks, called *task graph*.

Definition 1. A *Task Graph* (TG) $G = G(T, D)$ is a directed graph where each node represents a computational module in the application referred to as task $t_i \in T$. Each directed arc $d_{i,j} \in D$, between tasks t_i and t_j, characterizes either data or control dependencies.

Each task t_i is annotated with relevant information, such as a unique identifier and type of processing element (PE) in the network. Each $d_{i,j}$ is associated with a value $V(d_{i,j})$, which represents the volume of bits exchanged during the communication between tasks t_i and t_j.

Once the IP assignment has been completed, each task is associated with an IP identifier. The result of this stage is a graph of IPs representing the PEs responsible of executing the application.

Definition 2. An *Application Characterization Graph* (APG) $G = G(C, A)$ is a directed graph, where each vertex $c_i \in C$ represents a selected IP/core and each directed arc $a_{i,j}$ characterizes the communication process from core c_i to core c_j.

Each $a_{i,j}$ can be tagged with IP/application specific information, such as communication rate, communication bandwidth or a weight representing communication cost.

A TG is based on application features only while an APG is based on application and IP features, providing us with a much more realistic representation of the an application in runtime on a NoC. In order to be able to bind application and IP features, at least one common feature is required in both of the IP and TG models.

The E3S (0.9) Benchmark Suite [3] contains the characteristics of 17 embedded processors. These processors are characterized by the measured execution times of 47 different type of tasks, power consumption derived from processor datasheets, and additional information, such as die size, price, clock frequency and power consumption during idle state. In addition, E3S contains task graphs of common tasks in auto-industry, networking, telecommunication and office automation. Each one of the nodes of these task graphs is associated with a task type. A task type is a processor instruction or a set of instructions, e.g., FFT, inverse FFT, floating point operation, OSPF/Dijkstra [5], etc. If a given processor is able to execute a given type of instruction, so that processor is a candidate to receive a resource in the NoC structure and would be responsible for the execution of one or more tasks.

4.1 XML Representation

The E3S Benchmark Suite contains rich data about embedded processors and some common applications. TGFF [4], a random TG generator based on E3S processors, generates TGs with parallel and sequential tasks, nodes with IP types and other important features. Both, E3S and TGFF, are text files. We use XML Schema [20] to model the TG and IP repository. At this point, no standard schema for NoC design is available, so the XML structure for both representations reflects the features available from E3S processors and applications.

XML is a general-purpose specification for creating custom markup languages and we propose to use it as a standard in NoC design research. Its primary purpose is to help information systems share structured data and it is designed to be relatively human-legible. It is strongly structured and its structure can be controlled by a XSD schema. Any XML file based on a schema can be readable for any tool designed for that schema. To parse (process) a XML file is much easier than a TXT file. Modern programming languages offer APIs that facilitate XML files parsing, while TXT files must be read line by line, checking character by character. XML and XML schema (XSD) are defined by the World Wide Web Consortium (W3C) [20].

4.2 Task Graph Representation

Here, we represent TGs using the XML [20]. A TG is divided in three major elements: *taskgraph*, *nodes* and *edges*. The *taskgraph* element is the TG itself which contains *nodes* and *edges*. The *nodes* element includes a *node* element for each task of the TG and the *edges* element includes an *edge* element for each arc in the TG. Each *node* has two main attributes: an unique identifier (*id*) and a task type (*type*), chosen among the 47 different types of tasks present in the E3S. Each *edge* has four main attributes: an unique identifier (*id*), the *id* of its source node (*src*), the *id* of its target node (*tgt*) and an attribute representing the communication cost imposed (*cost*). Fig. 2 shows a simple TG and its corresponding XML representation.

4.3 IP Repository Representation

The IP repository is divided into two major elements: the *repository* and the *ips* elements. The *repository* is the IP repository itself. Recall that the repository contains different non general purpose embedded processors and each processor implements up to 47 different types of operations. Not all 47 different types of operations are available in all processors. Each type of operation available in each processor is represented by an *ip* element. Each *ip* is identified by its attribute *id*, which is unique, and by other attributes such as *taskType*, *taskName*, *taskPower*, *taskTime*,

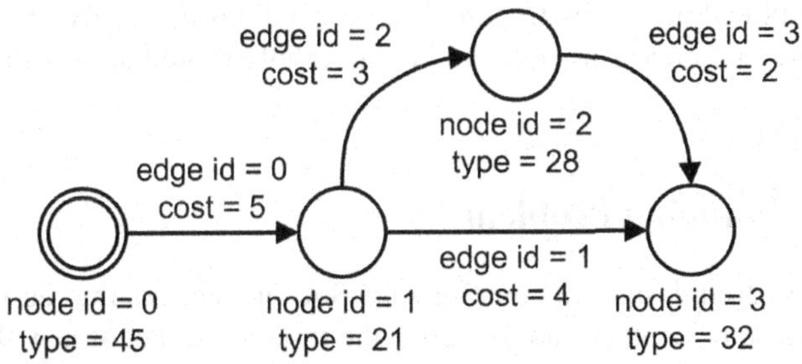


```
<?xml version="1.0" encoding="UTF-8"?>
<taskgraph>
   <nodes>
      <node id="0" type="45" .../>
      <node id="1" type="21" .../>
      <node id="2" type="28" .../>
      <node id="3" type="32" .../>
   </nodes>
   <edges>
      <edge id="0" src="0" tgt="1" cost="5"/>
      <edge id="1" src="1" tgt="3" cost="4"/>
      <edge id="2" src="1" tgt="2" cost="3"/>
      <edge id="3" src="2" tgt="3" cost="2"/>
   </edges>
</taskgraph>
```

Fig. 2 TG XML structure

processorID, *processorName*, *processorWidth*, *processorHeight*, *processorClock*, *processorIdlePower* and *cost*. The common element in TG and IP repository representations is the *type* attribute. Therefore, this element will be used to bind an *ip* to a *node*. Fig. 3 shows a simplified XML structure representing the IP repository. The repository contains IPs for digital signal processing, matrix operations, text processing and image manipulation.

```
<?xml version="1.0" encoding="UTF-8"?>
<repository>
   <ips>
      <ip id="10" type="0" procID="3" .../>
      <ip id="23" type="38" procID="5" .../>
      <ip id="68" type="12" procID="14" .../>
      <ip id="99" type="47" procID="17" .../>
   </ips>
</repository>
```

Fig. 3 IP repository XML structure

These simplified and well-structured representations are easily intelligible, improve information processing and can be universally shared among different NoC design tools.

5 The IP Mapping Problem

The platform-based design methodology for SoC encourages the reuse of components to increase reusability and to reduce the time-to-market of new designs. The designer of NoC-based systems faces two main problems: selecting the adequate set of IPs that optimize the execution of a given application and finding the best physical mapping of these IPs into the NoC structure.

The main objective of the IP assignment stage is to select, from the IP repository, a set of IPs that minimize the NoC consumption of power, area occupied and execution time. At this stage, no information about physical allocation of IPs is available so optimization must be done based on TG and IP information only. So, the result of this step is the set of IPs that maximizes the NoC performance. The TG is then annotated and an APG is produced, wherein each node has an IP associated with it.

Given an application, described by its APG, the problem that we are concerned with now is to determine how to topologically map the selected IPs onto the network, such that the objectives of interest are optimized. Some of these objectives are: latency requirements, power consumption of communication, total area occupied and thermal behavior. At this stage, a more accurate execution time can be calculated taking into account of the distance between resources and the number of switches and links crossed by a data package along a path. The result of this process should be an optimal allocation of the one of the prescribed IP assignments, selected in an earlier stage, to execute the application, described by the TG, on the NoC structure.

6 The Choice of Optimization Objectives

Different objectives may be considered in the IP mapping problem. If the improvement of an objective leads to the deterioration of another (e.g. maximizing clock frequency increases power consumption), the objectives are said to be *concurrent*. On the other hand, if the improvement of an objective leads to the improvement of another, the objectives are said to be *collaborative*. Optimization problems with *concurrent* and *collaborative* objectives are called Multi-objective Optimization Problems (MOPs). In such problems, all *collaborative* objectives should be grouped and a single objective among those should be used in the optimization process, which achieves also the optimization of all the collaborative objectives in the group. However, concurrent objectives need all to be considered in the optimization process. The best solution for a MOP is the solution with the adequate trade-off between *concurrent* and *collaborative* objectives.

Table 1 helps choosing the minimal set of objectives to be considered in IP mapping optimization stage. A up/down arrow in entry for objectives $i \times j$ means that

Table 1 Concurrent and Collaborative Objectives

	Area	Cost	Clock	Time	Power	# PEs
Area	↓	-	-	-	-	↓
Cost	-	↓	-	-	-	↓
Clock	-	-	↑	↓	↑	-
Time	-	-	↑	↓	↑	-
Power	-	-	↓	↑	↓	-
# PEs	↓	↓	-	-	-	↓

an increase/reduction with respect to objective i also leads to and increase/reduction with respect to objective j.

For instance, the last column of Table 1, which characterizes objective $\#PE$ (i.e. number of processor elements), indicates that a reduction with respect to this objective yields a reduction with respect to both *area* and *cost*. Therefore, those three objectives are considered collaborative. This is the same case for objective *time* and *clock frequency*. However, the penultimate column of Table 1, that characterizes objective *power*, indicates that a reduction in power leads to an increase in both time and clock frequency. Note that objective *power* must be minimized. Therefore, objective *power* is considered concurrent with both objective *time* and *clock frequency*. As a conclusion, the adequate trade-off can be achieved using only minimization functions of objectives *area*, *execution time* and *power consumption*.

7 Multi-objective Evolutionary Approach

The search space for a "good" IP mapping for a given application is defined by the possible combinations of IP/tile available in the NoC structure. Assuming that the mesh-based NoC structure has $N \times N$ titles and there are at most N^2 IPs to map, we have a domain size of $N^2!$. Among the huge number of solutions, it is possible to find many equally good solutions. In huge non-continuous search space, deterministic approaches do not deal very well with MOPs. The domination concept introduced by Pareto [16] is necessary to classify solutions. In order to deal with such a big search space and trade-offs offered by different solutions in a reasonable time, a multi-objective evolutionary approach is adopted.

The core of the proposed tool offers the utilization of two well-known and well-tested MOEAs: NSGA-II [2] and microGA [1]. Both adopt the domination concept with a ranking schema for classification. The ranking process separates solutions in *Pareto fronts* where each front corresponds to a given rank. Solutions from rank *one*, which is the *Pareto-optimal* front) are equally good and better than any other solution from Pareto fronts of higher ranks.

NSGA-II features a fast and elitist ranking process that minimizes computational complexity and provides a good spread of solutions. The elitist process consists in joining parents and offspring populations and diversity is achieved using the *crowded-comparison operator* [2]. MicroGA works with a very small population (3 to 5 individuals), which makes it very fast. A bigger population is stored on a population memory divided in replaceable and non-replaceable areas. The elitist process consists of storing the best solutions on a external memory and diversity is achieved using an *adaptive grid* [1].

The basic workflow of both algorithms is almost the same. They start with a random population of individuals, where each individual represents a solution. Each individual is associated with a rank. The selection operator is applied to select the parents. The parents pass through crossover and mutation operators to generate an offspring. A new population is created and the process is repeated until the stop criterion is satisfied.

7.1 Representation and Genetic Operators

The individual representation is shown in Fig. 4. The tile indicates information on the physical location on which a gene is mapped. On a $N \times N$ regular mesh, the tiles are numbered successively from top-left to bottom-right, row by row. The row of the ith tile is given by $\lceil i/N \rceil$, and the corresponding column by $i \bmod N$.

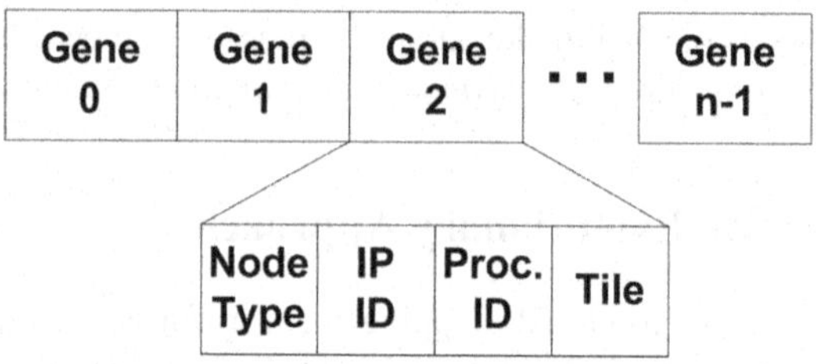

Fig. 4 Chromosome encoding of an IP mapping

The crossover and mutation operators were adapted to the fact that the set of selected IPs can not be changed as we have to adhere to the set of prescribed IP assignments. For this purpose, we propose a crossover operator that acts like a shift register, shifting around a random crossover point and so generating a new solution, but with the same set of IPs. This behavior does not contrast with the biological inspiration of evolutionary algorithms, observing that certain species can reproduce through parthenogenesis, a process in which only one individual is necessary to generate an offspring [19].

The mutation operator performs an *inner swap mutation*, where each gene receives a random mutation probability, which is compared against the system mutation probability. The genes with mutation probability higher than the system's are

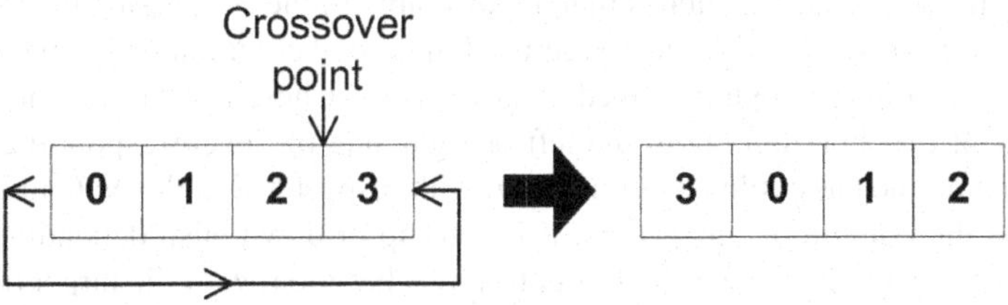

Fig. 5 Application of the proposed shift crossover

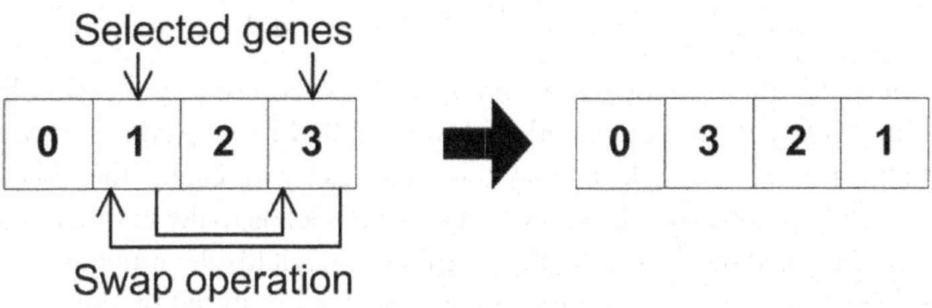

Fig. 6 Application of the proposed inner swap mutation

swapped with another random gene of the same individual, instead of selecting a random IP from the repository. This way, it is possible to explore the allocation space preserving any optimization done in the IP assignment stage. The crossover and mutation strategies adopted in the IP mapping stage are represented in Fig. 5 and Fig. 6, respectively.

8 Objective Function

During the evolutionary process, the fitness of the individuals with respect to each one of the selected objectives (i.e. *area*, *time*, and *power*) must be efficiently computed.

8.1 Area

In order to compute the area required by a given mapping of the application in question, it is necessary to know the area needed for the selected processors and that ocupied by the used links and switches. As a processor can be responsible for more than one task, each APG node must be visited in order to check the processor identification in the appropriate XML element. Grouping the nodes with the same *processorID* attribute allows us to implement this verification. The total number of links and switches can be obtained through the consideration of all communication paths between exploited tiles. Note that a given IP mapping may not use all the available tiles, links and switches that are available in the NoC structure. Also, observe that a portion of a path may be re-used in several communication paths.

In this work, we adopted a fixed route strategy wherein data emanating from tile i is sent first horizontally to the left or right side of the corresponding switch, depending on the target tile position, say j, with respect to i in the NoC mesh, until it reaches the column of tile j, then, it is sent up or down, also depending on the position of tile j with respect to tile i until it reaches the row of the target tile. Each communication path between tiles is stored in the routing table. The number of links in the aforementioned route can be computed as described in Equation 1. This is also represents the distance between tiles i and j and called the *Manhattan distance* [11].

$$nLinks(i, j) = |\lceil i/N \rceil - \lceil j/N \rceil| + |i \bmod N - j \bmod N| \quad (1)$$

In the purpose of computing efficiently the area required by all used links and switches, an APG can be associated with a so-called *routing table* whose entries describe appropriately the links and switches necessary to reach a tile from another. The number of hops between tiles along a given path leads to the number of links between those tiles, and incrementing that number by 1 yields the number of traversed switches. The area is computed summing up the areas required by the implementation of all distinct processors, switches and links.

Equation 2 describes the computation involved to obtain the total area for the implementation a given IP mapping M, wherein function *Proc*(.) provides the set of distinct processors used in APG$_M$ and $area_p$ is the required area for processor p, function $Links(.)$ gives the number of distinct links used in APG$_M$, A_l is the area of any given link and A_s is the area of any given switch.

$$Area(M) = \sum_{p \in Proc(APG_M)} area_p + (A_l + A_s) \times Links(APG_M) + A_s \quad (2)$$

8.2 Execution Time

To compute the execution time of a given mapping, we consider the execution time of each task of the critical path, their schedule and the additional time due to data transportation through links and switches along the communication path. The critical path can be found visiting all nodes of all possible paths in the task graph and recording the largest execution time of the so-called critical path. The execution time of each task is defined by the *taskTime* attribute in TG. Links and switches can be counted using the routing table. We identified three situations that can degrade the implementation performance, increasing the execution time of the application:

1. *Parallel tasks mapped into the same tile*: A TG can be viewed as a sequence of horizontal levels, wherein tasks to the same level may be executed in parallel (at the same time) allowing for a reduction of the overall execution time of the application. For instance, Figure 8 shows a TG that can be viewed as a sequence of 7 levels: $level_0 = \{t_0\}$; $level_1 = \{t_1, t_2\}$; $level_2 = \{t_3, t_4\}$;$level_3 = \{t_5, t_6, t_7, t_8\}$; $level_4 = \{t_9, t_{10}, t_{11}, t_{12}\}$; $level_5 = \{t_{13}, t_{14}\}$; and $level_6 =$

$\{t_{15}\}$. When parallel tasks are assigned in the same processor, which also means that these occupy the same tile of the NoC, they cannot be executed in parallel.
2. *Parallel tasks with partially shared communication path*: When a task in a tile (source) must send data to supposedly parallel tasks in different tiles (targets) through the same *initial* link, data to both tiles cannot be sent at the same time. For instance, using the mesh-base NoC of Figure 1, if the task on the most left upper tile have to send data to its right neighbor tile and to that on the most right upper tile at the same time, the initial link is common to both communication paths and so no parallelism can occur in such a case.
3. *Parallel tasks with common target using the same communication path*: When several tasks need to send data to a common target task, one or more shared links along the partially shared path would be needed simultaneously, the data from both tasks must then be pipelined and so will not arrive at the same time to the target task. For instance, using the mesh-base NoC if Figure 1, if the task on the most left upper tile and that on the most right upper tile have to send data to the center tile at the same time, they would send it to the right and left, respectively, and the upper center switch would buffer the data and send it in a pipelined manner to the center tile.

Equation 3 is computed using a recursive function that implements a depth-first search, wherein function $Paths(.)$ provides all possible paths of a given APG and $time_t$ is the required time for task *t*. After finding the including the total execution time of the tasks that are traversed by the critical path, the time of parallel tasks executed in the same processor need to be accumulated too. This is done by function *SameProcSameLevel*(.). The delay due to data pipelining for tasks on the same level is added by *SameSourceCommonPath*(.). Last but not least, the delay due to pipelining data that are emanating at the same time from several distinct tasks yet for the same target task is accounted for by function *DiffSrcSameTgt*(.).

$$
\begin{aligned}
Time(M) = \max_{r \in Paths(APG_M)} \Bigg(& \sum_{t \in r} time_t \\
& + SameProcSameLevel(r, TG, APG_M) \\
& + SameSrcCommonPath(r, TG, APG_M) \\
& + DiffSrcSameTgt(r, TG, APG_M) \Bigg)
\end{aligned} \tag{3}
$$

Function *SameProcSameLevel*(.) compares tasks of a given same level that are implemented by the same processor and returns the additional delay introduced in the execution of those tasks. Algorithm 1 shows how function $SameProcLevel(.)$, that uses information from path *r*, application task graph T and its corresponding characterization graph C to compute the delay in question.

Function *SameSourceCommonPath*(.) computes the additional time due to parallel tasks that have data dependencies on tasks mapped in the same source tile and yet these share a common initial link in the communication path. Algorithm 2

Algorithm 1. SameProcSameLevel(r, T, C)

```
time := 0
for all t ∈ r do
    for all n ∈ T do
        if T.level(t) = T.level(n) then
            if C.processor(t) = C.processor(n) then
                time := time + n.taskTime
return time
```

shows the details of the delay computation using information from path r, application task graph T and its corresponding characterization graph C. In that algorithm $T.targets(t)$ yields the list of all possible target tasks of task t, $A.initPath(src, tgt)$ returns the initial link of the communication path between tiles src and tgt and *penalty* represents a time duration needed to data to cross safely from one switch to one of its neighbors. This penalty is added every time the initial link is shared.

Algorithm 2. SameSrcCommonPath(r, T, C)

```
penalty := 0
for all t ∈ r do
    if T.targets(t) > 1 then
        for all n ∈ T.targets(t) do
            for all n' ∈ T.targets(t) | n' ≠ n do
                if C.initPath(t, n) = C.initPath(t, n') then
                    penalty := penalty + 1
return penalty
```

Function *DiffSrcSameTgt(.)* computes the additional time due to the fact that parallel tasks producing data for the same target task need to use simultaneously at least a common link along the communication path. Algorithm 3 shows the details of the delay computation using information from path r, application task graph T and its corresponding characterization graph C. In that algorithm, $C.Path(src, tgt)$ is an ordered list w of all links crossed through src task to tgt task and penalty has the same semantic as in the Algorithm 2.

8.3 Power Consumption

The total power consumption of an application NoC-based implementation consists of the power consumption of the processors while processing the computation performed by each IP and that due to the data transportation between the tiles. The former can be computed summing up attribute *taskPower* of all nodes of the APG and the latter is the power consumption due to communication between the application tasks through links and switches. The power consumption due to the computational activity is simply obtained summing up atribute *taskPower* of all nodes in the APG and is as described in Equation 4.

Algorithm 3. DiffSrcSameTgt(r, T, C)

```
penalty := 0
for all t ∈ r do
   for all t' ∈ r | t' ≠ t do
      if T.level(t) = T.level(t') then
         for all n ∈ T.targets(t) do
            for all n' ∈ T.targets(t') do
               if n = n' then
                  w := C.Path(t, n)
                  w' := C.Path(t', n')
                  for i = 0 to min(w.length, w'.length) do
                     if w(i) = w'(i) then
                        penalty := penalty + 1
return penalty
```

$$Power_p(M) = \sum_{t \in APG_M} power_t \tag{4}$$

The power consumption due to communication is a very important factor and must be considered in order to achieve low power consumption NoC designs. An energy model for one bit consumption is used to compute the total energy consumption for the whole communication involved during the execution of an application on the NoC platform. The bit energy (E_{bit}), energy consumed when a data of one bit is transported from one tile to any of its neighboring tiles, can be obtained as in Equation 5:

$$E_{bit} = E_{S_{bit}} + E_{L_{bit}} \tag{5}$$

wherein $E_{S_{bit}}$ and $E_{L_{bit}}$ represent the energy consumed by the switch and link tying the two neighboring tiles, respectively [7].

The total power consumption of sending one bit of data from tile i to tile j can be calculated considering the number of switches and links the bit passes through on its way along the path, as shown in Equation 6.

$$E_{bit}^{i,j} = nLinks(i,j) \times E_{L_{bit}} + (nLinks(i,j) + 1) \times E_{S_{bit}} \tag{6}$$

wherein function $nLinks(.)$ provides the number of traversed links (and switches too) considering the routing strategy used in this work and described earlier in this section. The function is is defined in Equation 1.

Recall that the application TG gives the communication volume ($V(t,t')$) in terms of number of bits sent from the task t to task t' passing through a direct arc $d_{t,t'}$. Assuming that the tasks t and t' have been mapped onto tiles i and j respectively, the communication volume of bits between tiles i and j is then $V(i,j)$ = $V(d_{t,t'})$. The communication between tiles i and j may consist of a single link $l_{i,j}$ or by a sequence of $m > 1$ links $l_{i,x_0}, l_{x_0,x_1}, l_{x_1,x_2}, \ldots, l_{x_{m-1},j}$. For instance, to establish a communication between tiles 0 and 8 (upper most left most and lower most right most) in the NoC structure of Figure 1, the sequence of links is therefore $l_{0,1}$, $l_{1,2}$, $l_{2,5}$ and $l_{5,8}$.

The total network communication power consumption for a given mapping M is given in Equation 7, wherein $Targets_t$ provides all tasks that have a direct dependency on data resulted from task t and $Tile_t$ yields the tile number into which task t is mapped.

$$Power_c(M) = \sum_{\substack{t \in APG_M, \\ \forall t' \in Targets_t}} V(d_{t,t'}) \times E_{bit}^{Tile_t, Tile_{t'}}, \tag{7}$$

Recall that the total power consumption of the application NoC-based implementation is given by the sum of the power consumption due the computational side of the application and that due to the communication involved between tiles, as explicitly shown in Equation 8.

$$Power(M) = Power_p(M) + Power_c(M) \tag{8}$$

9 Results

First of all, to validate the implementation of both algorithms, these were submitted to solve mathematical MOPs and the results were compared with their original results. Fig. 7 shows results of both algorithms for a two objective optimization problem called KUR, proposed by Kursawe and used by Deb and Coello to validate NSGA-II [2] and microGA [1], respectively.

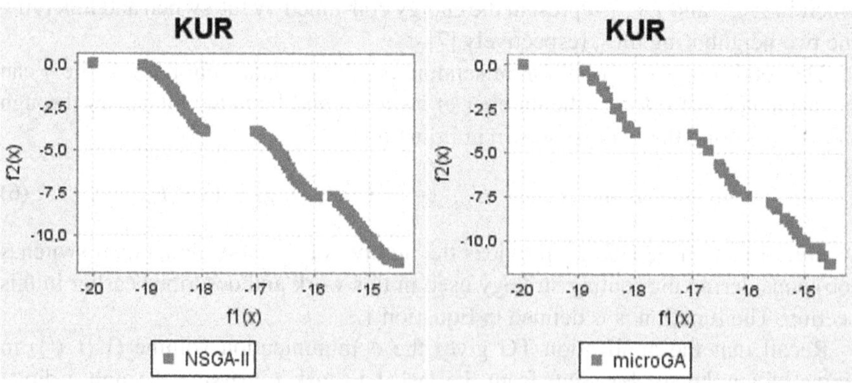

Fig. 7 Results for KUR function

Both algorithms converged to the true Pareto-front. As expected, NSGA-II found a higher diversity of solutions while microGA was much more faster. The parameters used for these tests were the same used by their original authors.

For NoC optimization, only the individual representation and the objective functions were changed, keeping the ranking, selection, crossover and mutation operators unchanged. Different TGs generated with TGFF [4] and from E3S, with sequential and parallel tasks, were used.

Many simulations were performed to find out the setting up of the parameters used in NSGA-II and micro-GA. For the former, we used a population size of 600, mutation probability of 0.01, crossover probability of 0.8 and tournament size of 50 and run the algorithm of 100 generations. For the latter, we used population memory size of 1000, replaceable fraction of 0.7, non-replaceable fraction of 0.3, micro population of 4 individuals, mutation probability of 0.02, crossover probability of 0.09, tournament size of 2, external memory of 200, nominal convergence of 4, replacement cycle of 100, bisection of 5, and run the algorithm for 3000 generations.

The application, represented as a TG in Fig. 8, was generated with TGFF [4]. This TG presents five levels of parallelism, formed by the mirrored left-right side nodes.

Analyzing the results obtained from the first simulations, we observed that in order to achieve the best trade-off, the system allocated many tasks for the same

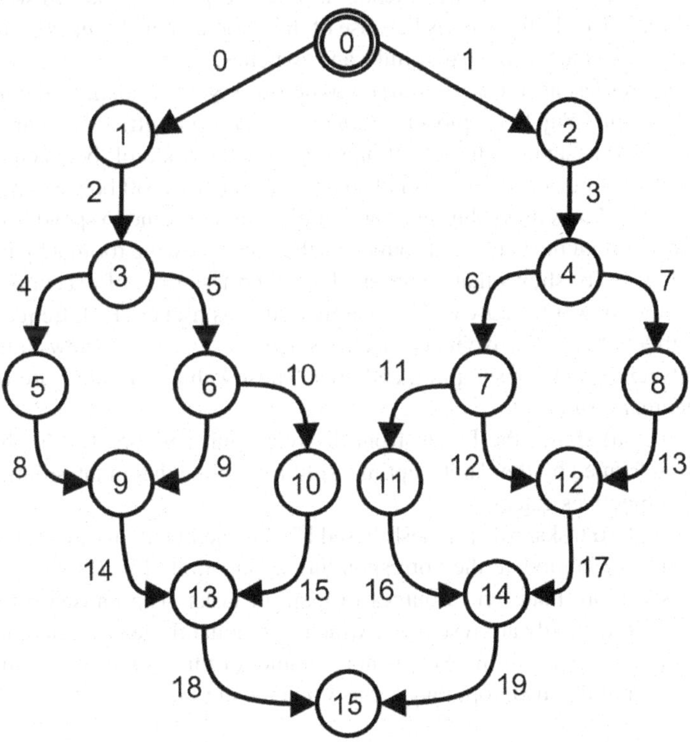

Fig. 8 Task graph with five levels of parallelism

processor, which reduces area and execution time but generates *hot spots* [21]. A hot spot is an area of high activity within a silicon chip. Hot spots can damage a silicon chip and increases power consumption because of Avalanche Effect; an effect that increases reverse current in semiconductor materials because of temperature rising and other factors [12]. In order to avoid the formation of hot spots, a *maximum tasks per processor* constraint was imposed in the evolutionary process. This parameter is decided by the NoC designer based on some extra physical characteristics. We adopted a maximum of 2 tasks per processor.

The IP assignment [17] of the TG represented in Fig. 8 was able to discover 97 distinct optimal IP assignments. From those 97 distinct of IP assignments, 142 optimal mappings were generated. The 15 most significant solutions from the Pareto-front, with respect to each of the considered objectives, are listed in Table 2 and Table 3. Table 2 presents the IPs assigned to each node of the TG and the respective fitness in terms of to each of the selected objectives after running the IP assignment step. Table 3 presents the tile where each assigned IP was mapped and Table 4 shows the respective fitness in terms of each of the selected objective after completing the IP mapping step. The first five solutions impose shorter execution times; the next five require smaller hardware areas and the last five solutions present lower power consumptions. The differences in execution time and power, observed when comparing the data from both tables is because of the inflicted penalty in execution time and power due to the use of shared links and switches.

Fig. 9–(a) represents the $time \times area$ trade-off, Fig. 9–(b) depicts the $power \times time$ trade-off and Fig. 9–(c) plots the $power \times area$ trade-off. As we can observe, comparing the dots against the line of interpolation, the trade-off between time and area and between power and time is not so linear as the trade-off between power and area. Fig. 9–(a) shows that solutions that require more area tend to spend less execution time because of the better distribution of the tasks allowing for more parallelism to occur. Fig. 9–(b) shows that solutions that spend less time of execution tend to consume more power because of IP's features, such as higher clock frequency, and physical effects like intensive inner-electrons activity. Fig. 9–(c) shows a linear relation between power consumption and area. Those values and units are based on E3S Benchmark Suite [3].

Figure 10–(a) shows the Pareto-front discrete points. Figure 10–(b) shows the Pareto-front formed by the Pareto-optimal solutions. Note that many solutions have very close objectives values.

For a TG of 16 tasks, a 4×4 mesh-based NoC is the maximal physical structure necessary to accommodate the corresponding application. Table 4 shows that no solution used more than ten resources to map all tasks. The unused 6 tiles may denote a waste of hardware resources, which consequently lead to the conclusion that either the geometry of the NoC is not suitable for this application or the mesh-based NoC is not the ideal topology for its implementation.

Table 2 Optimal IP assignments for the task graph of Fig. 8

solution	IP/Task - From task 0 to task 15	time (s)	area[1]	power (W)
1	[942, 242, 938, 858, 380, 523, 720, 0, 855, 66, 379, 13, 544, 721, 661, 239]	0.1224	19.1014	52.95
2	[942, 457, 938, 378, 860, 523, 720, 0, 855, 494, 379, 13, 879, 721, 244, 239]	0.1289	5.777	21.125
3	[942, 937, 458, 378, 860, 523, 240, 0, 375, 14, 859, 493, 879, 721, 724, 239]	0.1324	5.777	21.125
4	[462, 907, 458, 858, 10, 43, 720, 480, 375, 494, 379, 880, 862, 721, 244, 239]	0.1439	5.6271	23.55
5	[942, 937, 458, 858, 10, 523, 720, 867, 375, 864, 379, 13, 214, 456, 724, 239]	0.1502	5.777	19.6
6	[369, 457, 938, 378, 860, 367, 935, 480, 207, 696, 859, 383, 492, 456, 724, 239]	0.2072	2.9976	15.84
7	[849, 457, 938, 378, 490, 847, 455, 480, 855, 216, 379, 13, 862, 936, 42, 239]	0.2582	3.239	15.04
8	[942, 937, 458, 378, 490, 43, 240, 480, 855, 216, 379, 13, 862, 456, 724, 719]	0.1502	3.8586	18.7
9	[942, 457, 938, 8, 860, 367, 720, 370, 485, 14, 859, 383, 492, 456, 244, 719]	0.1828	3.9586	16.87
10	[462, 937, 458, 858, 10, 847, 240, 370, 855, 384, 9, 493, 694, 936, 724, 239]	0.1818	3.9586	17.27
11	[369, 457, 938, 378, 490, 367, 935, 0, 375, 14, 859, 493, 862, 456, 244, 239]	0.2072	3.239	15.04
12	[369, 457, 938, 378, 10, 367, 935, 850, 375, 216, 859, 13, 694, 456, 724, 239]	0.2072	2.9976	15.84
13	[462, 457, 938, 858, 490, 367, 240, 0, 375, 864, 379, 13, 492, 936, 724, 719]	0.1818	3.9586	16.87
14	[462, 457, 938, 858, 490, 847, 240, 0, 375, 384, 859, 13, 694, 936, 244, 719]	0.1818	3.9586	17.27
15	[462, 937, 458, 210, 860, 43, 240, 480, 375, 384, 859, 13, 492, 936, 724, 719]	0.1502	3.8586	18.7

[1] (x$10^{-5}m^2$).

Table 3 Optimal IP mappings for the task graph of Fig. 8

solution	IPs / Tiles - From $tile_0$ to $tile_{15}$
1	[–, (380, 379), –, –, (0, 13), (720, 721), 66, –, 661, 554, (242, 239), –, 523, (858, 855), (942, 938), –]
2	[–, (0, 13), –, –, –, (244, 239), (378, 379), –, 457, (523, 494), (720, 721), –, (942, 938), 879, (860, 855), –]
3	[458, (942, 937), (860, 859), –, –, (721, 724), (378, 375), (879), –, (0, 14), (240, 239), –, –, (523, 493), –, –]
4	[(375, 379), (858, 862), (720, 721), –, –, (244, 239), (480, 494), –, (10, 43), (462, 458), (907, 880), –, –, –, –, –]
5	[–, –, 867, (10, 13), –, (858, 864), (214, 239), (720, 724), –, 523, (942, 937), (458, 456), –, –, –, (375, 379)]
6	[–, –, –, –, (369, 367), (378, 383), (696, 724), –, (457, 456), (860, 859), (207, 239), –, –, (480, 492), (938, 935), –]
7	[–, –, –, –, –, –, (855, 862), (490, 480), –, (457, 455), (216, 239), (13, 42), –, (378, 379), (849, 847), (938, 936)]
8	[(942, 937), (240, 216), –, –, (458, 456), (378, 379), (43, 13), –, (724, 719), (855, 862), (490, 480), –, –, –, –, –]
9	[(942, 938), (860, 859), (370, 383), –, (720, 719), (485, 492), 244, –, (8, 14), (457, 456), –, –, 367, –, –, –]
10	[–, –, –, –, –, (694, 724), 847, (858, 855), –, (370, 384), 493, (240, 239), (462, 458), (937, 936), –, (10, 9)]
11	[–, –, –, –, (378, 375), –, (490, 493), (859, 862), (369, 367), (244, 239), –, –, (457, 456), (938, 935), (0, 14)]
12	[(216, 239), (938, 935), (457, 456), –, –, (378, 375), (369, 367), (694, 724), –, (10, 13), (850, 859), –, –, –, –, –]
13	[(0, 13), (724, 719), (490, 492), (375, 379), –, –, –, –, –, (462, 457), (938, 936), –, –, 367, (858, 864), 240]
14	[–, –, –, –, –, (462, 457), (858, 859), 847, –, 490, (938, 936), (375, 384), –, (0, 13), (240, 244), (694, 719)]
15	[–, –, (462, 458), –, (43, 13), (937, 936), (860, 859), (724, 719), (375, 384), (210, 240), (480, 492), –, –, –, –, –]

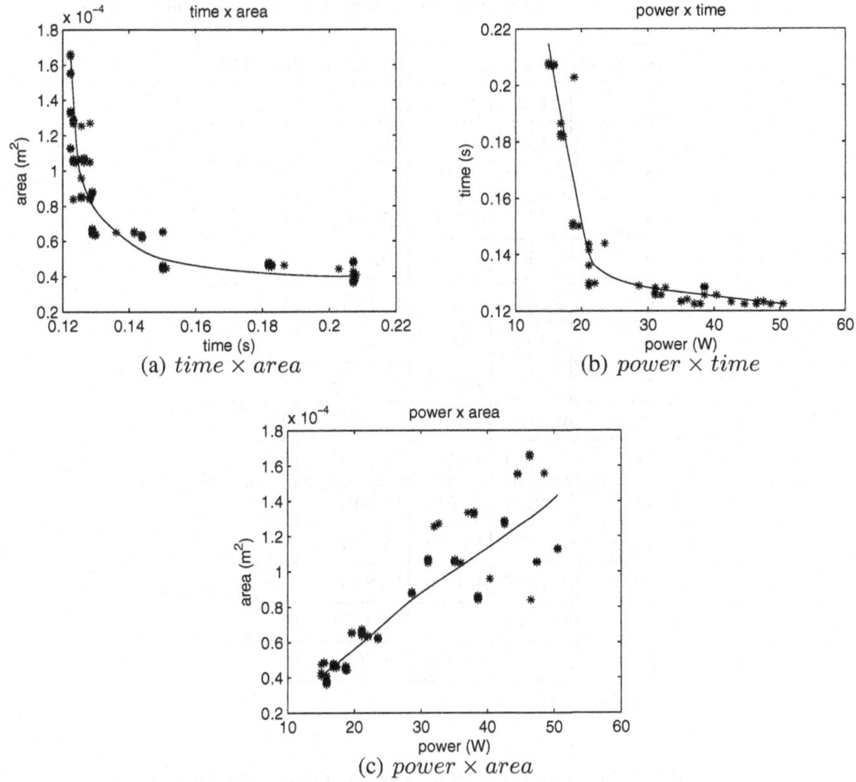

(a) $time \times area$

(b) $power \times time$

(c) $power \times area$

Fig. 9 Trade-offs representation of the 142 optimal IP mappings obtained for the task graph of Fig. 8

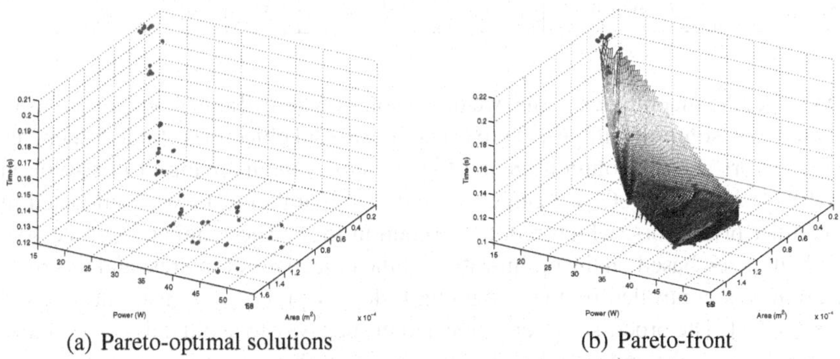

(a) Pareto-optimal solutions

(b) Pareto-front

Fig. 10 Pareto-optimal solutions and Pareto-front

Table 4 Characteristics of optimal IP mappings for the task graph of Fig. 8

solution	time (s)	area[1]	power (W)	Used tiles
1	0.1225	19.9514	52.9501	10
2	0.129	6.386	21.1251	9
3	0.1362	6.506	21.1251	9
4	0.144	6.2951	23.5501	8
5	0.1503	6.566	19.6001	9
6	0.2073	3.6056	15.8401	8
7	0.2583	3.787	15.0401	8
8	0.1503	4.4066	18.7001	8
9	0.1829	4.5676	16.8701	9
10	0.1819	4.6276	17.2701	9
11	0.2073	4.267	15.0401	8
12	0.2073	3.7856	15.8401	8
13	0.1819	4.8076	16.8701	9
14	0.1819	4.6276	17.2701	9
15	0.1503	4.6466	18.7001	8

[1] (x$10^{-5}m^2$).

Table 5 Processors of solution #8 from Table 2 and Table 3

TG Node	0	1	2	3	4	5	6	7	8	9	10	11	12	13	14	15
Proc ID	32	32	15	13	17	0	6	17	30	6	13	0	30	15	23	23
IP ID	942	937	458	378	490	43	240	480	855	216	379	13	862	456	724	719
Tile	0	0	4	5	10	6	1	10	9	1	5	6	9	4	8	8

As a specific mapping example, we detail solution #8 in Table 3, which seems to be a moderate solution with respect to every considered objectives. Table 5 specifies the processors used in solution #8 and Fig. 11 shows the mapping of this solution into the mesh-based NoC. We can observe that all parallel tasks were allocated in the distinct processors, which reduces execution time. The number of processors were minimized based on the optimization of the objectives of interest and this minimization was controlled by the maximum tasks per processor constraint to avoid hot spots [21]. The processors were allocated in such way to avoid delay of communication due to links and switches disputed by more than one resource at the same time.

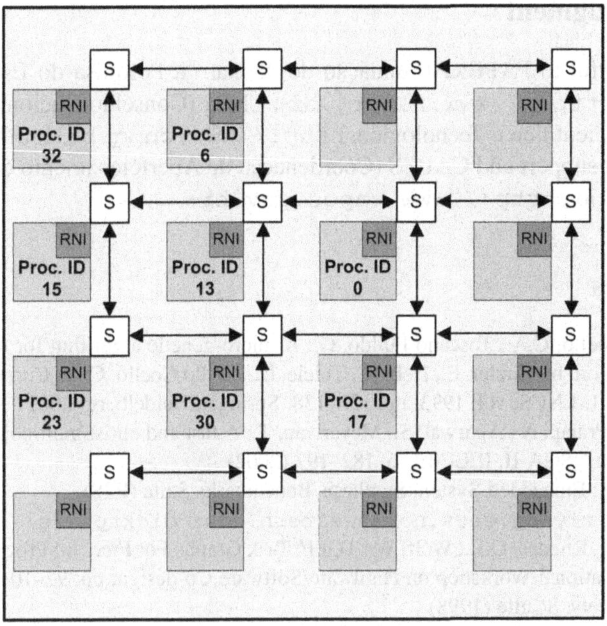

Fig. 11 Mapping of solution #8 from Table 2 and Table 3

10 Conclusions

The problem of mapping IPs into a NoC topology is NP-hard [6] key research problems in NoC design [15]. In this paper we propose a decision support system based on MOEAs to help NoC designers allocate a prescribed set of IPs into a NoC physical structure. The use of two different MOEAs consolidates the obtained results.

Structured and intelligible representations of a NoC, a TG and of a repository of IPs were used and these can be easily extended to different NoC applications. Despite of the fact that we have adopted E3S Benchmark Suite [3] as our repository of IPs, any other repository could be used and modeled using XML, making this tool compatible with different repositories.

The proposed *shift crossover* and *inner swap mutation* genetic operators can be used in any optimization problem where no lost of data from a individual is accepted.

Future work is three-fold: adopting a dynamic topology strategy to try to evolve the most adequate topology for a given application; exploring the use of different objectives based on different repositories and proposing an interfacing mechanism with a hardware description simulator to integrate our tool to the NoC design platform.

Acknowledgment

We are grateful to FAPERJ (Fundação de Amparo à Pesquisa do Estado do Rio de Janeiro, `http:// www.faperj.br`), CNPq (Conselho Nacional de Desenvolvimento Científico e Tecnológico, `http://www.cnpq.br`) for their continuous financial support and CAPES (Coordenação de Aperfeiçoamento de Pessoal de Ensino Superior, `http://www.capes.gov.br`).

References

1. Coello Coello, C.A., Toscano Pulido, G.: A micro-genetic algorithm for multiobjective optimization. In: Zitzler, E., Deb, K., Thiele, L., Coello Coello, C.A., Corne, D.W. (eds.) EMO 2001. LNCS, vol. 1993, pp. 126–138. Springer, Heidelberg (2001)
2. Deb, K., Pratap, A., Agarwal, S., Meyarivan, T.: A fast and elitist multiobjective genetic algorithm: NSGA-II. IEEE-EC 6, 182–197 (2002)
3. Dick, R.P.: Embedded System Synthesis Benchmarks Suite (E3S), `http://ziyang.eecs.northwestern.edu/~dickrp/e3s/`
4. Dick, R.P., Rhodes, D.L., Wolf, W.: TGFF: Task Graphs For Free. In: Proceedings of the 6th International Workshop on Hardware/Software Co-design, pp. 97–101. IEEE Computer Society, Seattle (1998)
5. Dijkstra, E.W.: A note on two problems in connexion with graphs. Numerische Mathematik 1, 269–271 (1959)
6. Garey, M.R., Johnson, D.S.: Computers and intractability; a guide to the theory of *NP*-completeness. Freeman and Company, New York (1979)
7. Hu, J., Marculescu, R.: Energy-aware mapping for tile-based NoC architectures under performance constraints. In: ASPDAC: Proceedings of the 2003 Conference on Asia South Pacific Design Automation, Kitakyushu, Japan, pp. 233–239. ACM, New York (2003), doi:10.1145/1119772.1119818
8. ISO/IEC. ISO/IEC 7498-1:1994: Information technology – Open Systems Interconnection – Basic Reference Model: The Basic Model. ISO, Geneva, Switzerland, 1994.
9. Jena, R.K., Sharma, G.K.: A multi-objective evolutionary algorithm based optimization model for network-on-chip synthesis. In: ITNG, pp. 977–982. IEEE Computer Society, Los Alamitos (2007)
10. Kumar, S., Jantsch, A., Millberg, M., Öberg, J., Soininen, J.-P., Forsell, M., Tiensyrjä, K., Hemani, A.: A network on chip architecture and design methodology. In: ISVLSI, pp. 117–124. IEEE Computer Society, Los Alamitos (2002)
11. Lei, T., Kumar, S.: A two-step genetic algorithm for mapping task graphs to a network on chip architecture. In: DSD, pp. 180–189. IEEE Computer Society, Los Alamitos (2003)
12. Lu, P.-F., Chen, T.-C.: Collector-base junction avalanche effects in advanced double-poly self-aligned bipolar transistors. IEEE Transactions on Electron Devices 36(6), 1182–1188 (1989)
13. Murali, S., De Micheli, G.: Bandwidth-constrained mapping of cores onto NoC architectures. In: DATE, pp. 896–903. IEEE Computer Society, Los Alamitos (2004)
14. Murali, S., De Micheli, G.: SUNMAP: a tool for automatic topology selection and generation for nocs. In: Proceedings of the 41st Annual conference on Design Automation (DAC 2004), June 7-11, pp. 914–919. ACM Press, New York (2004)

15. Ogras, Y., Hu, J., Marculescu, R.: Key research problems in NoC design: a holistic perspective. In: Eles, P., Jantsch, A., Bergamaschi, R.A. (eds.) Proceedings of the 3rd IEEE/ACM/IFIP International Conference on Hardware/Software Codesign and System Synthesis, CODES+ISSS 2005, pp. 69–74. ACM, New York (2005)
16. Pareto, V.: Cours D'Economie Politique. F. Rouge, Lausanne (1896)
17. da Silva, M.V.C., Nedjah, N., Mourelle, L.M.: Evolutionary IP assignment for efficient noc-based system design using multi-objective optimization. In: Proceedings of the Congress of Evolutionary Computation (CEC 2009), Norway, May 18-21, IEEE Press, Los Alamitos (2009), page (submitted)
18. Srinivas, N., Deb, K.: Multiobjective function optimization using nondominated sorting in genetic algorithms. Evolutionary Computation 2(3), 221–248 (1995)
19. Watts, P.C., Buley, K.R., Sanderson, S., Boardman, W., Ciofi, C., Gibson, R.: Parthenogenesis in komodo dragons. Nature 444(7122), 1021–1022 (2006)
20. World Wide Web Consortium (W3C) (2008), `http://www.w3.org`
21. Zhou, W., Zhang, Y., Mao, Z.: Pareto based multi-objective mapping IP cores onto NoC architectures. In: APCCAS, pp. 331–334. IEEE, Los Alamitos (2006)

Theory and Applications of Chaotic Optimization Methods

Tohru Ikeguchi[1], Mikio Hasegawa[2], Takayuki Kimura[3], Takafumi Matsuura[4], and Kazuyuki Aihara[5]

[1] Department of Information and Computer Sciences, Saitama University, 255 Shimo-ohkubo, Sakura-ku, Saimata, 338-8570, Japan
tohru@mail.saitama-u.ac.jp
[2] Department of Electrical Engineering, Faculty of Engineering, Tokyo Univesity of Science, 1-14-6 Kudankita, Chiyoda-ku, Tokyo 102-0073, Japan
hasegawa@ee.kagu.tus.ac.jp
[3] Department of Electrical and Electronic Engineering, Faculty of Engineering, Nagasaki University, 1-14 Bunkyo-machi, Nagasaki, 852-8521, Japan
k1mura@nagasaki-u.ac.jp
[4] Department of Management Science, Faculty of Engineering, Tokyo Univesity of Science, 1-3 Kagurazaka, Shinjuku-ku, Tokyo 162-8601, Japan
matsuura@ms.kagu.tus.ac.jp
[5] Institute of Industrial Science, University of Tokyo, 4-6-1 Komaba, Meguro-ku, Tokyo 153-8505, Japan
aihara@sat.t.u-tokyo.ac.jp

1 Introduction

In our society, various combinatorial optimization problems exist and we must often solve them, for e.g. scheduling, delivery planning, circuit design, and computer wiring. Then, one of the important issues in science and engineering is how to develop effective algorithms for solving these combinatorial problems.

To develop effective algorithms for such combinatorial optimization problems in the real world, the standard combinatorial optimization problems are intensively studied: for example, traveling salesman problems, quadratic assignment problems, vehicle routing problems, packet routing problems, and motif extraction problems. Among them, the traveling salesman problem (TSP) is one of the most standard combinatorial optimization problems . The TSP is described as follows: when a set of N cities and distances d_{ij} between cities i and j are given, find an optimal solution, or the shortest-length tour visiting all the cities once. Namely, the goal of TSP is to find a permutation $\boldsymbol{\sigma}$ of the cities that minimizes the following objective function:

$$L(\boldsymbol{\sigma}) = \sum_{k=1}^{N-1} d_{\sigma(k)\sigma(k+1)} + d_{\sigma(N)\sigma(1)}, \tag{1}$$

N. Nedjah et al. (Eds.): Innovative Computing Methods, SCI 357, pp. 131–161.
springerlink.com

where $L(\boldsymbol{\sigma})$ is the tour length of the TSP with $\boldsymbol{\sigma}$. If $d_{ij} = d_{ji}$ for all i and j, the TSP is symmetric; otherwise, it is asymmetric. In this chapter, we deal with the symmetric TSP.

For an N-city symmetric TSP, the number of all possible tours is $(N-1)!/2$. Thus, the number of tours factorially diverges if the number of cities increases. It is widely acknowledged that the TSP belongs to a class of NP-hard. Therefore, it is required to develop an effective approximate algorithm for finding near-optimal solutions in a reasonable time frame.

To discover approximate solutions, various heuristic algorithms have already been proposed. In 1985, Hopfield and Tank proposed an approach for solving TSPs by using a recurrent neural network. They applied descent downhill dynamics of the recurrent neural network to obtain approximation solutions of TSPs [1]. Although this approach is interesting from the viewpoint of an application of neural dynamics to real engineering problems, such as combinatorial optimization, it has two drawbacks.

The first drawback is that this approach has a local minimum problem: it uses simple descent downhill dynamics of the neural network to obtain better solutions of TSPs; states of the neural networks can be stuck at undesirable local minima. To resolve the local minimum problems, two main strategies that uses chaotic dynamics have been proposed. The first solution is to inject chaotic noise into the dynamics of the neural network [2,3,4,5,6]. The second solution is to replace the descend downhill dynamics with chaotic dynamics [7, 8, 9, 10]. Recently, these strategies have been applied to combinatorial optimization in engineering, for example, channel assignment problems [11], frequency assignment problems [12], multicast routing problems [13], and broadcast routing problems [13]. The performance of using chaotic dynamics shows that the algorithm finds an optimal or near-optimal solutions of the problems. In this chapter, we review basic theories of these strategies in Sects. 2.1 and 2.2.

Although the first drawback can be resolved by the above-mentioned strategies, namely chaotic noise injection and chaotic dynamics in recurrent neural networks, there still exists the second drawback. The methods based on the recurrent neural networks cannot be applied to large scale instances of combinatorial optimization, because it takes a huge amount of memories to construct the neural network. In addition, it is not so easy to obtain a feasible solution. In the method, a solution of TSPs is encoded by a firing pattern of the neural network. Thus, a solution is generated only in the case that the firing pattern of the neural network completely satisfies a constraint.

To resolve the second drawback, an approach in which heuristic algorithms are controlled by the chaotic dynamics has been proposed [14, 15, 16, 17, 18]. In this approach with chaotic dynamics, to avoid local minima, execution of a heuristic algorithm, such as the 2-opt algorithm , the 3-opt algorithm , the Or-opt algorithm [19] , the Lin-Kernighan (LK) algorithm [20] , and the stem-and-cycle (S&C) ejection chain method [21,22], is controlled by chaotic dynamics.

In [14, 15, 16, 17, 18], to generate the chaotic dynamics, a chaotic neural network [23, 24] is used. In the chaotic neural network, the basic element is a chaotic neuron proposed by Aihara et al. [23, 24]. The model introduced two important properties which real nerve cells have, namely graded response and refractoriness . The latter means that when a neuron has just fired, the firing of this neuron is inhibited for a while by the refractoriness.

In [14,15,16,17,18], execution of the local search algorithm is encoded by firing of the chaotic neuron. If the chaotic neuron fires, the corresponding local search algorithm is executed. Because the firing of the chaotic neuron is inhibited by the refractoriness, frequent firing of the chaotic neuron, or frequent execution of the local search algorithm is restricted. This can be a mechanism for the chaotic search to escape from local minima efficiently. It is reported that the refractoriness implemented in the chaotic neuron model leads to a higher solving ability than the tabu search which has a similar strategy of searching solutions as the chaotic search [15].

On the basis of the above idea that the refractoriness of the chaotic neuron can be used for an effective escape from local minima, a chaotic search method which controls the 2-opt algorithm has already been proposed [14, 15]. Although the 2-opt algorithm is the simplest local search algorithm, the chaotic search method with the 2-opt algorithm shows good results. In [14], in the case of solving an N-city TSP, $N \times N$ chaotic neurons are prepared and arranged on $N \times N$ grid. Here, $N \times N$ neurons correspond to the number of possible ways for constructing a new tour by the 2-opt improvement. As a result, this chaotic search method shows solving performance with less than 0.2% gaps from the optimal solution for instances with the order of 10^2 cities [14].

On the other hand, the tabu search [25,26] is also a quite effective strategy for escaping from local minima. The chaotic search method in [15] is based on the tabu search which is implemented on a neural network, because both the tabu search and the chaotic search have similar strategies that forbid backward moves. In [15], by assigning one neuron to one city, only N chaotic neurons are used to solve an N-city TSP. As a result, this chaotic search method can be applied to large scale examples, such as the 85,900-city one, and it has high solving performance with less than 3.0% gaps from the optimal solutions for instances with the order of 10^4 cities [15]. In addition, this strategy has been improved by introducing other local searches, such as the Or-opt algorithm [19], the Lin-Kernighan (LK) algorithm [20], and the stem-and-cycle (S&C) ejection chain method [21, 22]. In [27, 28, 29, 30, 31, 17, 32], these sophisticated local searches are controlled by chaotic dynamics. The results show that large scale TSPs could be solved by these chaotic searches.

In this chapter, the chaotic search that resolves the second drawback in the Hopfield-Tank neural network approach is described in In 3, we also review several applications of this chaotic search to real engineering problems: the vehicle routing problems in 3.1, the motif extraction problems in 3.2, and the packet routing problems in 3.3.

2 Methods

In this section, we review the methods for solving combinatorial optimization problems by chaotic dynamics.

2.1 Chaotic Noise

Performance improvement by the chaotic noise

Effectiveness of the chaotic noise for shaking the states of the solution search has been shown by many papers [2,3,4,5,6]. In Figs. 1 and 2, the solvable performances of the Hopfield-Tank neural networks [1] with stochastic noise and chaotic noise are compared by applying to the TSP and QAP, respectively. For these comparisons, the logistic map chaos, $z(t+1) = az(t)(1 - z(t))$, is introduced as a most simple example of chaotic sequence, with the parameters $a = 3.82$, $a = 3.92$, and $a = 3.95$. As the stochastic noise, the Gaussian white noise is introduced. The horizontal axis is noise amplitude β and the vertical axis is the percentage of the optimum solution obtained by 1,000 different initial conditions. From the figures, it is clear that the chaotic noise is effective for combinatorial optimization algorithm using the recurrent neural networks. Although the original Hopfield-Tank neural network quickly converges to a stable state corresponding to a local optimal and does not offer the optimum solution for these problems in any cases, its performance can be much improved by adding the chaotic noise, which makes the solvable performances almost 100% for the 20-city TSP as shown in Fig. 1 and around 20% for the 12-node QAP as shown in Fig. 2, respectively. The white Gaussian noise also improves the performance of the Hopfield-Tank neural network, but its solvable performances are around 80% and 6% for the TSP and the QAP, respectively, and much lower than the chaotic noise cases. These results show that the chaotic noise has much better solvable performance than the stochastic noise.

The Hopfield-Tank neural network can be applied to combinatorial optimization problems, based on its minimization property of the energy function

$$E(t) = -\frac{1}{2}\sum_{i=1}^{n}\sum_{j=1}^{n}\sum_{k=1}\sum_{l=1} w_{ikjl}x_{ik}(t)x_{jl}(t) + \sum_{i=1}^{n}\sum_{k=1}^{n} \theta_{ik}x_{ik}(t), \qquad (2)$$

which always decreases by asynchronous update of each neuron by the following simple equation:

$$x_{ik}(t+1) = f\left[\sum_{j=1}^{n}\sum_{l=1}^{n} w_{ikjl}x_{jl}(t) + \theta_{ik}\right], \qquad (3)$$

where $x_{ik}(t)$ is the output of the ikth neuron at time t, w_{ikjl} is the connection weight between the ikth and jlth neurons, and θ_{ik} is the threshold of the ikth neuron.

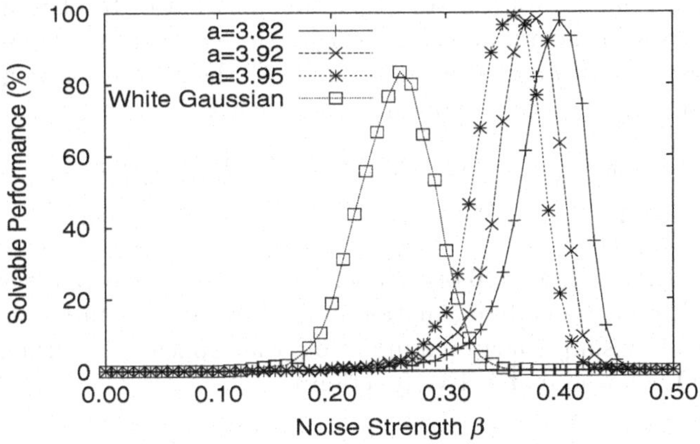

Fig. 1 Solvable performance of the recurrent neural networks with the chaotic noise and the white Gaussian noise on a 20-city TSP

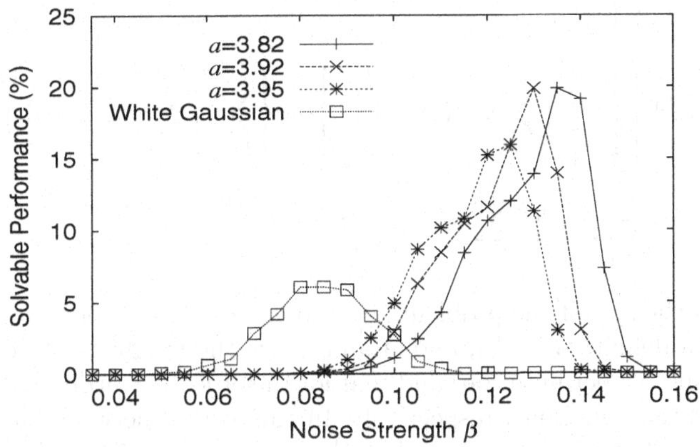

Fig. 2 Solvable performance of the recurrent neural networks with the chaotic noise and the white Gaussian noise on a QAP (Nug12)

As already been applied in many previous researches, the energy function for solving the TSPs [1, 7, 8, 10, 2, 3] can be defined by the following equation:

$$E_{\text{TSP}} = A\left[\left\{\sum_{i=1}^{N}\left(\sum_{k=1}^{N}x_{ik}(t)-1\right)^2\right\}+\left\{\sum_{k=1}^{N}\left(\sum_{i=1}^{N}x_{ik}(t)-1\right)^2\right\}\right] + B\sum_{i=1}^{N}\sum_{j=1}^{N}\sum_{k=1}^{N}d_{ij}x_{ik}(t)\{x_{jk+1}(t)+x_{jk-1}(t)\}, \quad (4)$$

where N is the number of cities, d_{ij} is the distance between the cities i and j, and A and B are the weight of the constraint term (formation of a closed tour) and that for the objective term (minimization of total tour length), respectively. In this neural network, firing of the ikth neuron means that the city i is visited at the k order. From (2) and (4), the connection weights w_{ijkl} and the threshold θ_{ijkl} can be obtained as follows:

$$w_{ikjl} = -A\{\delta_{ij}(1-\delta_{kl}) + \delta_{kl}(1-\delta ij)\} - Bd_{ij}(\delta_{lk+1} + \delta_{l-k-1}), \tag{5}$$

$$\theta_{ij} = 2A, \tag{6}$$

where $\delta_{ij} = 1$ if $i = j$, $\delta_{ij} = 0$ otherwise. Using these connection weights and the thresholds for the update equation in (3), the better solution of the TSP may be found by simple neuronal updating, that appears as a firing pattern.

For the QAPs whose objective function is

$$F(\mathrm{p}) = \sum_{i=1}^{N}\sum_{j=1}^{N} a_{ij} b_{p(i)p(j)}, \tag{7}$$

which has to be minimized by finding the optimal permutation $\mathbf{p}$, we use the following energy function:

$$E_{\mathrm{QAP}} = A\left[\left\{\sum_{i=1}^{N}\left(\sum_{k=1}^{N} x_{ik}(t) - 1\right)^2\right\} + \left\{\sum_{k=1}^{N}\left(\sum_{i=1}^{N} x_{ik}(t) - 1\right)^2\right\}\right]$$
$$+ B\sum_{i=1}^{N}\sum_{j=1}^{N}\sum_{k=1}^{N}\sum_{l=1}^{N} a_{ij} b_{kl} x_{ik}(t) x_{jl}(t), \tag{8}$$

where N is the size of the problem, a_{ij} and b_{kl} are $N \times N$ matrices given in the problem definition in (7), and A and B are the weight of the constraint term (making $\mathbf{p}$ a permutation) and that for the objective term (minimization of the objective function), respectively. In this neural network formulation, firing of the ikth neuron means that the element i is assigned to the kth location of the permutation. By transforming (8) to the form of the energy function of the recurrent neural network in (2), the connection weights and the thresholds for solving the QAPs are obtained as follows:

$$w_{ikjl} = -A\{\delta_{ij}(1-\delta_{kl}) + \delta_{kl}(1-\delta_{ij})\} - Ba_{ij}b_{kl}, \tag{9}$$

$$\theta_{ij} = 2A. \tag{10}$$

Using these connection weights and the thresholds for neuronal updates by (3), the better solutions of the QAP may be obtained.

However, the Hopfield-Tank neural network is well known to have the local minimum problem, because the energy function simply decreases and it can search simply one solution in a huge number of local optimal solutions. To improve the performance of such a neural network, the chaotic noise or stochastic noise has been applied to this optimization neural network to improve the solutions by avoiding trapping at the undesirable optimal states.

In [3,4,6], the following update equation is used to apply the noise to the neural network,

$$x_{ik}(t+1) = f\left[\sum_{j=1}^{N}\sum_{l=1}^{N} w_{ikjl}x_{jl}(t) + \theta_{ik} + \beta z_{ik}(t)\right], \tag{11}$$

where $z_{ik}(t)$ is a noise sequence added to the ikth neuron, β is the amplitude of noise, and f is the sigmoidal output function, $f(y) = 1/(1+\exp(-y/\epsilon))$, respectively. The noise sequence introduced as $z_{ik}(t)$ should be normalized to zero mean and unit variance.

The results in Figs. 1 and 2 are obtained by using the neural network with noise described above, with changing the noise amplitude β. From the results, the noise amplitude β for the best performance differs among the noise. By comparing the best solvable performances in each noise, effectiveness of the chaotic noise can be clearly shown by such simple experiments.

Analysis on the effects of the chaotic noise

In order to know why the chaotic noise is more effective than other noise, such as the Gaussian white noise, effectiveness of the chaotic noise has been analyzed from various aspects.

In [2], Hayakawa et al. compared the performance of the neural networks with the chaotic noise generated by the logistic map and those with randomly shuffled sequences of the chaotic noise, whose temporal structure, such as auto-correlation, is broken. Their results show that the performance with the random shuffled sequence becomes much worse than the original chaotic sequence. From such results, they anticipated that the temporal structure of the chaotic noise is important in the chaotic search.

In [3], Hasegawa et al. presented much clearer results showing the importance of the temporal structure of the chaotic noise, by applying the method of surrogate data [33]. They introduced three algorithms for generating surrogate data. The first one is the random shuffle algorithm, which is almost the same as the method which was introduced in [2] mentioned above. It preserves the empirical histogram of the original data. The second one is the Fourier transformed surrogate algorithm, which generates stochastic data preserving the auto-correlation function and the power spectrum of the original data. Such surrogate data can be generated by applying the discrete Fourier transform to the original data for obtaining the power spectrum, randomizing the phase with keeping the same power spectrum, and then applying the inverse discrete Fourier transform to the phase randomized spectrum. The generated sequence will have the same power spectrum and auto-correlation function as the original data. The third algorithm also preserves the auto-correlation and power spectrum of the original data, and additionally the empirical histogram of the original data.

The results in [3] show that the neural networks with the noise sequences generated by the second and the third surrogate algorithms, which preserve the auto-correlation function and the power spectrum of the original chaotic sequence, have almost the same performances as those of the neural network with the original chaotic sequence. This result clearly shows that temporal structure, such as auto-correlation function, of each noise is an important factor for high performance of the chaotic search.

In [4], Hasegawa and Umeno focused on the shape of the auto-correlation function of the chaotic sequences which leads the neural network with high performance. Such chaotic sequences have the auto-correlation that gradually converges to zero with damped oscillation. Such chaotic sequences with similar auto-correlation have also been utilized in the chaotic CDMA researches [34,35]. In CDMA, minimization of the cross-correlation among the spreading sequences leads to the lower mutual interference. In the chaotic CDMA researches such as [34,35], the chaotic sequences, whose auto-correlation are $C(\tau) \approx c \times (-r)^{\tau}$, have been used, that is similar to the chaotic sequences which leads high performance on the optimization neural network described above, where r is the damping factor, τ is the lag, and c is a constant. It has been shown that the auto-correlation with the $r = 2 - \sqrt{3}$ leads to the minimum cross-correlation among the sequences. By using such a sequences, performance of the bit error rate in the CDMA communication could be improved in [35].

Hasegawa and Umeno also investigated the performance of the neural network with such noise whose auto-correlation is $C(\tau) \approx c \times (-r)^{\tau}$, and showed that higher performance can be realized only by the noise with positive r which is similar auto-correlation as the original chaotic sequence. Furthermore, in [6], Minami and Hasegawa showed that injection of negative auto-correlation sequences leads to the lower cross-correlation that may be realized by the same mechanism as the chaotic CDMA [34,35]. From these researches, it has been shown that the neural networks with the chaotic noise have higher solving abilities because their negative auto-correlation makes smallest cross-correlation between the neurons that leads the high dimensional searching dynamics of the neural network to the most complicated dynamics, and such dynamics makes the performance much higher than other noise sequences such as the white noise.

2.2 Recurrent Chaotic Neural Networks

Performance improvement by the chaotic neural networks

The chaotic neural network [23,24,36] has an inherent property to escape from any fixed points except that of all the resting neurons due to accumulated refractory effects. This property was first applied to dynamical associative memory [23, 37], then to the optimization methods based on the Hopfield-Tank neural networks [1], and effectiveness of chaotic dynamical searches has also been shown by many papers [7,8,9,10]. The chaotic neural network is a neural network model consisting of the chaotic neurons, which have the chaotic dynamics. The chaotic neuron is an extension of the Nagumo-Sato binary neuron model, which has an exponentially decreasing refractory effect, to an analog version by replacing the Heaviside output function to the sigmoidal output function. The chaotic neuron map can be described as follows:

$$y(t+1) = ky(t) - \alpha f(y(t)) + a, \tag{12}$$

where $y(t)$ is the internal state of the neuron at time t, k is the decay parameter of the refractory effects, α is the scaling parameter of the refractory effects, a is a positive bias, and f is a sigmoidal function, $f(y) = 1/(1 + \exp(-y/\epsilon))$. The chaotic dynamics can be represented by this bimodal nonlinear map [24,36].

The chaotic neural network is a network composed of the chaotic neurons, which is defined as follows [23,36,37]:

$$\eta_{ij}(t+1) = k_m \eta_{ij}(t) + \sum_k \sum_l W_{ijkl} x_{kl}(t), \tag{13}$$

$$\zeta_{ij}(t+1) = k_r \zeta_{ij}(t) - \alpha x_{ij}(t) + a_{ij}, \tag{14}$$

$$x_{ij}(t+1) = f[\eta_{ij}(t+1) + \zeta_{ij}(t+1)], \tag{15}$$

where $\eta_{ij}(t)$ is the internal state for mutual connections of the ijth neuron at time t, $\zeta_{ij}(t)$ is the internal state for the refractory effects of the ijth neuron at time t, a_{ij} is the positive bias for the ijth neuron, k_m and k_r are the decay parameters for the mutual connections and refractory effects, respectively.

In [7,8,9], the chaotic neural network with a single internal state which corresponds to setting $k_r = k_m$ is used. Effectiveness of such chaotic dynamics has been shown by Nozawa in [7], by comparing the performances with those of the stochastic models on the basis of extension of the continuous-time Hopfield neural network model to a discrete-time model with adding negative self-feedback connections for each neuron. Such a negative self-feedback connection corresponds to the refractory effects in the above chaotic neural network model, and this neural network has the chaotic dynamics as well. To improve the performance of the chaotic search, Chen and Aihara proposed chaotic simulated annealing by gradually reducing the effects of the chaotic fluctuation in the searching dynamics [8], and showed that the performance can be much improved.

Analysis on the effects of the chaotic neural network

In the chaotic noise approach in the previous section, the chaotic dynamics has been introduced as additive noise to the gradient dynamics. In contrast with such an approach, the chaotic neural network approach uses the searching dynamics whose dynamics itself is chaotic. Therefore, the searching dynamics of the chaotic neural networks has various well-known characteristics of the chaotic dynamics, such as orbital instability , self-similarity , and so on, and has better performance than the chaotic noise approach [38].

Such chaotic searching dynamics of the chaotic neural networks has been analyzed by calculating the Lyapunov exponents [9, 10]. Although it is not easy to estimate the Lyapunov exponents accurately for such high dimensional chaotic dynamical systems, clear results have been obtained. Yamada and Aihara calculated the Lyapunov exponents of the single internal state model [9]. They analyzed the performance of the chaotic neural network with changing its parameter values, and showed that sum of the positive Lyapunov exponents of the high performance chaotic dynamics becomes close to zero. They argued that such results mean that the edge of chaos , between the periodic dynamics and the chaotic dynamics, has the best performance for combinatorial optimization. Hasegawa et al. also analyzed the relation between the solvable performances of the chaotic neural networks and its Lyapunov exponents on the two-internal-state model described in (13) and (14). They showed that it is possible to tune the dynamics of the chaotic neural network by changing the balance between two decay parameters, k_r and k_m, for the internal states $\eta_{ij}(t)$ and $\zeta_{ij}(t)$, respectively, and obtained similar results that the chaotic dynamics with smaller positive Lyapunov exponents has the best performance. They also calculated the coefficient of variation, as a complexity index of the firing interval of the neurons, and showed that the chaotic dynamics with small Lyapunov exponents has higher complexity that makes the chaotic neural network realize high solvable performances.

From the obtained results in various researches on the both approaches using chaotic dynamics, the chaotic noise and the chaotic neural networks, for the recurrent neural networks, described in the previous and present sections, it has been understood that chaos makes the searching dynamics very complicated and the performance improved. Such chaotic dynamics is more complex than the stochastic dynamics in a sense, and the better performance can be realized. By further researches, it may be possible to completely clarify the mechanism of the effectiveness of the chaotic search in the recurrent neural network in the near future.

2.3 Chaotic Search That Controls Execution of Heuristic Algorithm

Recurrent chaotic neural networks are effective for solving combinatorial optimization problems as shown in the previous section 2.2, However, the method cannot be applied to very large instances because it needs huge amounts of memory to construct the neural network. In addition, it is not so easy to obtain feasible solutions because a firing pattern of a chaotic neural network encodes a solution. Thus, a solution is generated only in the case that the firing pattern of the neural network satisfies the constraints. To resolve these serious problems, the second approach, in which heuristic algorithms are driven by the chaotic dynamics, has been proposed [14].

In the approach, execution of a local search algorithm is controlled by the chaotic dynamics. The basic element is a chaotic neuron proposed by Aihara et al. [23, 24]. Execution of the local search algorithm is encoded by firing of the chaotic neuron. If the chaotic neuron fires, the corresponding local search algorithm is executed. After a neuron has just fired, the next firing of this neuron is inhibited for a while by the refractoriness of the chaotic neuron. Thus, frequent firing of the chaotic neuron, or frequent execution of the local search algorithm is restricted. Therefore, the chaotic search can escape from local minima efficiently. Then, it is reported that the refractoriness implemented in the chaotic neuron model leads to equivalent or even higher solving ability than tabu search which has almost the same strategy of searching solutions as the chaotic search.

Using the above-mentioned idea, chaotic search methods have already been proposed to find the near optimal solutions or approximate solutions for combinatorial optimization problems such as traveling salesman problems [14,39,40,15,16,18,27,17,32], quadratic assignment problems [41,42], vehicle routing problems [43,44,45], motif extraction problems [46,47,48,49,18], and packet routing problems [50, 51, 52, 53, 54, 55, 56, 57, 58]. In this section, we describe the simplest chaotic search method for solving TSP [14, 39, 40, 15]. In the method, execution of the 2-opt algorithm is driven by the chaotic neurodynamics . The 2-opt algorithm exchanges two paths with other two paths until no further improvement can be obtained (Fig. 3). However, a tour obtained by the 2-opt algorithm is not a global optimum but a local optimum. To jump from such a local optimum, we applied the chaotic neurodynamics to the 2-opt algorithm [14,41,15]. To realize the chaotic search and to avoid local minima by the chaotic dynamics, a chaotic neuron is assigned to each city. Then, if a chaotic neuron fires, the local searches related to the corresponding city are carried out.

Dynamics of the ith chaotic neuron is defined as follows:

$$x_i(t+1) = f(y_i(t+1)), \tag{16}$$

$$f(y) = \frac{1}{1+\exp(-y/\epsilon))}, \tag{17}$$

$$y_i(t+1) = \xi_i(t+1) + \zeta_i(t+1), \tag{18}$$

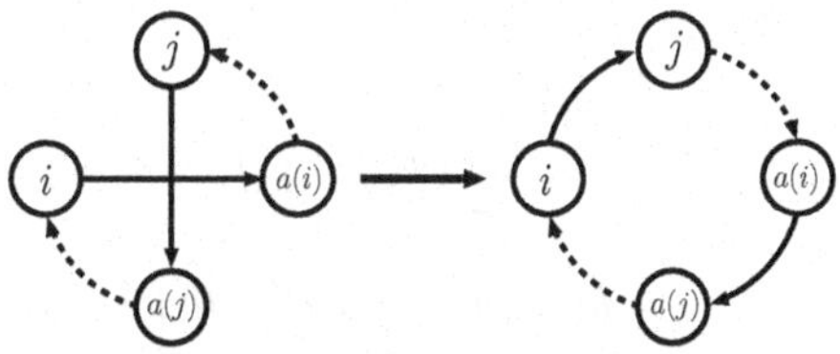

Fig. 3 An example of the 2-opt algorithm. In this example, $a(i)$ is the next city to i. Two paths (i-$a(i)$ and j-$a(j)$) are deleted from the current tour, then new two paths, i-j and $a(j)$-$a(i)$, are added to obtain a shorter tour

where $x_i(t+1)$ is an output of the ith chaotic neuron at time $t+1$; $f(\cdot)$ is a sigmoidal function; $y_i(t+1)$ is an internal state of the ith chaotic neuron at time $t+1$. The internal state is decomposed into two effects: a gain effect and a refractory effect.

The gain effect is defined by

$$\xi_i(t+1) = \max_j\{\beta(t)\Delta_{ij}(t) + \zeta_j(t)\}, \tag{19}$$

where $\Delta_{ij}(t)$ is difference between the length of a current tour and that of a new tour in which city j is next of city i after applying the 2-opt algorithm to city i (Fig. 3). $\zeta_j(t)$ is a refractory effect of the jth city at time t which is defined in (21). In (19), the refractory effect of the jth city is included to avoid local minima. Let us assume that a searching state now gets stuck at a local minimum. Then, we calculate a maximum value of Δ_{ij}. In this case, the maximum value is $\Delta_{ij} = 0$ because the current solution is a local optimal solution or an optimal solution. Thus, to select other cities, the refractory effect is included in (19). In this equation, $\beta(t)$ is a scaling parameter, which increases in proportion to time t as follows:

$$\beta(t+1) = \beta(t) + \lambda. \tag{20}$$

If we use (20), the searching space is gradually limited, which has a similar characteristic as the simulated annealing [59]. If $\beta(t)$ takes a small value, the method can explore a large solution space. On the other hand, if $\beta(t)$ takes a large value, the method works like a greedy algorithm.

The refractory effect works to avoid the local minima, which has a similar effect as a memory in the tabu search [25, 26]. In the tabu search, to avoid a local minimum, previous states are memorized by adding them to a tabu list and are not allowed again for a certain temporal duration called a tabu tenure. In the case of the chaotic search, past firing is memorized as previous states to decide the present strength of the refractory effect, which increases just after corresponding neuron firing and recovers exponentially with time. Thus, while the tabu search perfectly inhibits to select the same solutions for the certain period, the chaotic search might permit to select the same solutions if a corresponding neuron fires due to a larger gain effect or an exponential decay of the refractory effect. The refractory effect is expressed as follows:

$$\zeta_i(t+1) = -\alpha \sum_{d=0}^{t} k_r^d x_i(t-d) + \theta \tag{21}$$

$$= k_r\zeta_i(t) - \alpha x_i(t) + \theta(1-k_r), \tag{22}$$

where α controls the strength of the refractory effect after the firing ($0 < \alpha$); the parameter k_r is the decay parameter of the refractory effect ($0 < k_r < 1$); and θ is the threshold value. If the neuron frequently fires in its past history, the first term of the right hand side of (35) becomes negative. Then the neuron leads to a resting state. By using the chaotic neuron, we can realize the tabu search method as a special case [25, 26, 41, 15]. The conventional tabu effect can be described by setting the parameters $\alpha \to \infty$ and $k_r = 1$ in the refractory effect ζ_i' of the chaotic neuron as follows:

$$\zeta_i' = -\alpha \sum_{d=0}^{s-1} k_r^d x_{ij}(t-d) + \theta, \tag{23}$$

where s corresponds to a tabu tenure. If the neuron fired during s steps, firing of this neuron is inhibited by the parameters α and k_r. It is considered that the computational cost is almost the same as the tabu search method. We have already shown that the chaotic search method shows better performance than the tabu search method in several examples [41, 15, 42, 47, 43, 44, 50, 51, 52].

When we solve the TSP, the following procedure is used:

1. An initial tour is constructed, for example by the nearest neighbor method.
2. The tour is improved by the 2-opt algorithm controlled by chaotic dynamics.
   a. A city i is selected from the neurons whose internal state has not been updated yet.
   b. A city j is selected in such a way that the gain effect is maximum.
   c. If the ith neuron fires, city i and city j are connected by the 2-opt algorithm.
   d. The steps a)-c) are repeated until all neurons are updated.
3. One iteration is finished. Then, the step 2 is repeated.

To extend the above-mentioned algorithm, we have also proposed a new chaotic search method [17]. In the method, execution of the LK algorithm [20] is controlled by the chaotic dynamics. The LK algorithm [20] is one of the most powerful variable depth search methods. It can explore better solutions than the adaptive k-opt algorithm because the adaptive k-opt algorithm is based on a simpler rule. The number of exchanged links k increases when a gain of a $(k+1)$-opt improvement is larger than that of a k-opt improvement. As a result, the chaotic search method using the LK algorithm shows solving performance with less than 0.7% gaps from the optimal solution for instances with the order of 10^4 cities and can be applied to large scale instances with the order of 10^5 cities [17].

On the other hand, the S&C ejection chain method [21, 60] is also one of the most effective variable depth search methods. It is reported that the S&C ejection chain method leads to better solutions than the LK algorithm [60]. One of the reasons is that the S&C ejection chain method can explore more diversified solution space, because it introduces an S&C structure, which is not a tour. Namely, the S&C ejection chain method can explore infeasible

Table 1 The results of the local search algorithms and the chaotic search methods. Results are expressed by percentages of gaps between obtained solutions and the optimal solutions (%)

Instance	Local search 2-opt	2opt + Or-opt [19]	LK [20]	SC [21]	Chaotic search 2-opt [14,39,40,15]	2opt+ Or-opt [27]	LK [17]	SC [32]
pcb442	7.473	3.970	2.298	2.148	1.034	0.409	0.336	0.369
pcb1173	9.885	6.238	2.903	2.680	1.692	0.804	0.599	0.452
pr2392	9.563	5.294	4.225	3.252	1.952	1.153	0.619	0.466
rl5915	9.395	6.244	4.311	3.268	2.395	1.291	0.748	0.702
rl11849	8.752	5.567	4.066	3.248	2.223	1.139	0.708	0.652

solution space. However, the S&C ejection chain method also gets stuck at local minima because it is also a greedy algorithm.

Table 1 shows performances of chaotic search methods for benchmark instances of TSPLIB [61] (see [27,17,32] for details of the algorithms). Although the computational cost of the chaotic search methods is larger than that of the local search method because the refractory effect is calculated to avoid local minima, the performance of the local search methods is much improved by the chaotic dynamics.

3 Applications

In the previous section, we described the three basic approaches for solving combinatorial optimization problems with chaotic dynamics. In this section, we review the application of the chaotic search methods to three important engineering applications: vehicle routing problems, motif extraction problems, and packet routing problems.

3.1 Vehicle Routing Problems

To plan an efficient schedule of the delivery, the vehicle routing problem (VRP) is widely studied [62,25,26,63,64,43,44,45]. In this section, we explain a chaotic search method for solving VRP. Then, we show the results for the Solomon's benchmark problems [65] and the Gehring and Homberger benchmark problems [66].

The VRP consists of a depot, vehicles and customers. The depot is a departure and an arrival point of the vehicles. Each vehicle has a weight limit and visits the customers to satisfy their demands. The customers are visited only once by one vehicle. Then, the object of the VRP is to minimize the number of vehicles and the total travel distance. Generally speaking, a primary object of the VRP is to minimize of the number of vehicles. Thus, we use the following objective function:

$$g(S) = \sum_{l=1}^{m} D_l + \gamma \times m, \tag{24}$$

where S is a solution (a set of tours of all vehicles), m is the number of vehicles, D_l is the total travel distance of the lth vehicle, and γ is the scaling parameter. Because the first priority of the VRP is to reduce the number of vehicles, we set γ large. We deal with a VRP with time windows (VRPTW). In the VRPTW, each customer has its own time window, and the vehicles have to visit the customers within the time window. If the time windows are violated, solutions are infeasible.

In the chaotic search method, we use two simple local searches. The first one is to exchange the customer for another one, and the second one is to relocate the customer to another place. In the method of [44, 43, 45], $2n$ neurons are needed to solve an n-customer problem. Each neuron corresponds to each customer. If a neuron fires, a customer corresponding to the neuron is exchanged or relocated.

To realize the chaotic search, each neuron has a gain effect and a refractory effect. These effects of the ijth neuron are defined as follows:

$$\xi_{ij}(t+1) = \max_{s}\{\beta \Delta_{ijs}\}, \tag{25}$$

$$\zeta_{ij}(t+1) = -\alpha \sum_{d=0}^{t} k_r^d x_{ij}(t-d) + \theta, \tag{26}$$

where $\xi_{ij}(t)$ and $\zeta_{ij}(t)$ represent the gain effect and the refractory effect, respectively. Then, an output of the ijth neuron is defined as follows:

$$x_{ij}(t+1) = f\{\xi_{ij}(t+1) + \zeta_{ij}(t+1)\}, \tag{27}$$

where $f(y) = 1/(1+e^{-y/\epsilon})$. If $x_{ij}(t) > 1/2$, the ijth neuron fires at time t, and the local search to which the neuron corresponds is performed.

In (25), β is the positive scaling parameter of the gain effect, and Δ_{ijs} is the gain value of the objective function (24) if the local searches are performed. $\Delta_{ijs} = g(S_B) - g(S_A)$, where S_B and S_A are solutions before and after the local searches are performed respectively. Here, s indicates a customer to be exchanged (Fig. 4(a)) or relocated to their next order (Fig. 4(b)), and s is so selected that Δ_{ijs} takes the maximum gain. By the gain effect, the neurons corresponding to good operations become easy to fire.

In (26), α is the positive scaling parameter, $k_r (0 < k_r < 1)$ is the decay factor, and θ is the threshold value. Then, the refractory effect inhibits firing of the neuron which has just fired; this realizes a memory effect with an exponential decay. The strength of the refractory effect gradually decays depending on the value of k_r.

In the method of [44, 43, 45], the neurons are asynchronously updated. When all the neurons are updated, a single iteration is finished.

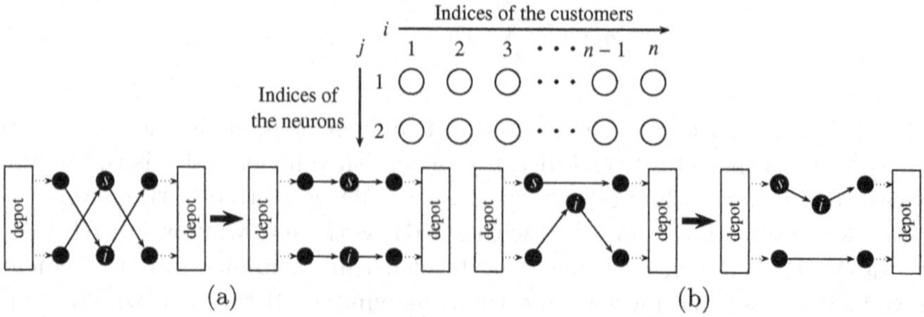

Fig. 4 Construction of chaotic neurons and two local searches used in the chaotic search method: (**a**) Exchange of one customer for another customer and (**b**) Relocation of one customer for another place. If the ijth neuron fires, (**a**) the ith customer is exchanged ($j = 1$) for another one customer s or (**b**) the ith customer is relocated ($j = 2$) to a next order of customer s

To evaluate the performance of the proposed method, we solved the Solomon's benchmark problems [65] and the Gehring and Homberger benchmark problems [66]. We produced an initial solution using the Bräysy construction heuristic method [67]. Then, we treated time windows as hard constraints. The parameters in (25)–(27) are set as follows: $\beta = 0.04, \alpha = 0.5, k_r = 0.9, \theta = 0.9$, and $\epsilon = 0.02$.

Results are shown in Table 2. The simulations are conducted by an Intel Core 2 Duo 2GHz computer for 1,000 iterations. Table 2 shows the average numbers of vehicles and the average total travel distances (italic) for each problem type. These results show that the proposed method provides good results. In addition, we compared the proposed method with the other conventional methods (see [64, 43, 44, 45] for details). The results indicated that the proposed method has higher performance than the other conventional methods by changing the values of parameter effectively depending on the constraints of each problem.

3.2 Motif Extraction Problems

To identify an important region in biological sequences, the motif extraction problem (MEP) is solved in bioinformatics. In this section, we explain a Chaotic Motif Sampler (CMS), that employs chaotic dynamics to solve the MEPs [68, 46, 47, 48, 69, 49, 70].

The definition of the MEP can be mathematically described as follows [71]: we have a biological data set $S = \{s_1, s_2, ..., s_N\}$, where s_i is the ith sequence (Fig. 5). Each sequence consists of m_i ($i = 1, 2, ..., N$) elements. In the case of DNA or RNA, the elements correspond to four bases; while in the case of protein sequences, they correspond to 20 amino acids. $V = \{v_1, v_2, \cdots, v_N\}$ is a set of motifs, where the length of the motif is L (Fig. 5). Of course, the

Table 2 Results for the 100-customer instances from the Solomon's benchmark problems [65] and the other customer instances from the Gehring and Homberger benchmark problems [66]. In this table, R means a random allocation, C a clustered allocation, and RC their mixture

type	100	200	400	600	800	1000
R1	12.6	18.4	36.7	55.4	73.0	92.9
	1230.5	*3865.4*	*9166.1*	*19274.1*	*33496.4*	*50210.7*
R2	3.1	4.1	8.1	11.1	15.1	19.2
	961.3	*3180.8*	*6768.3*	*13978.5*	*22639.3*	*33250.0*
C1	10.0	19.2	38.4	58.1	76.7	96.1
	840.2	*2755.9*	*7337.2*	*14278.7*	*25310.8*	*42376.3*
C2	3.0	6.0	12.1	18.2	24.2	30.3
	592.5	*1878.9*	*4102.5*	*7887.0*	*12196.9*	*17563.1*
RC1	12.0	18.4	36.8	55.4	73.3	90.8
	1388.1	*3446.5*	*8486.4*	*17453.5*	*30521.0*	*47038.2*
RC2	3.4	4.7	9.4	12.6	16.8	19.4
	1195.2	*2719.4*	*5654.3*	*11519.1*	*18114.2*	*27952.0*

alignment of the motifs and the length of the motifs are unknown. Then, the aim of the MEP is to find a set of motifs that maximizes the following objective function:

$$E = \frac{1}{L} \sum_{k=1}^{L} \sum_{\omega \in \Omega} f_k(\omega) \log_2 \frac{f_k(\omega)}{p(\omega)}, \tag{28}$$

where $f_k(\omega)$ is the appearance frequency of an element (a base in the case of DNA or RNA sequences and an amino acid in the case of a protein sequences) $\omega \in \Omega$ at the kth position of motif candidates; Ω is a set of bases or a set of amino acids; and $p(\omega)$ is the background probability of appearance of the element ω (Fig. 5). In (28), $0 \log_2 0$ is defined to be 0.

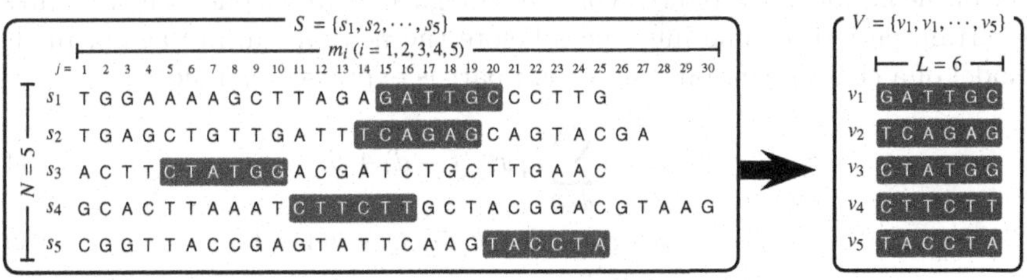

Fig. 5 Definition of the motif extraction problem

In the CMS, to extract motifs $v_1, v_2, ..., v_N$, a chaotic neuron is assigned to the head positions of all motif candidates (Fig. 6). The firing of the ijth chaotic neuron is then defined as follows:

$$x_{ij}(t) = f(y_{ij}(t)) > \frac{1}{2}, \tag{29}$$

$$f(y) = \frac{1}{1 + \exp(-y/\epsilon)}, \tag{30}$$

$$y_{ij}(t) = \xi_{ij}(t) + \zeta_{ij}(t), \tag{31}$$

where $x_{ij}(t)$ is an output state of the ijth chaotic neuron at time t. If the chaotic neuron fires or $x_{ij}(t+1) > \frac{1}{2}$, the ijth motif position becomes a motif candidate. On the other hand, if the ijth neuron is resting or $x_{ij}(t+1) \leq \frac{1}{2}$, the ijth motif candidate is not selected. $y_{ij}(t)$ is an internal state of the ijth chaotic neuron at time t. The internal state of the chaotic neuron [23] is decomposed into two parts. The first part $\xi_{ij}(t)$ represents the gain effect of the ijth neuron at time t and the second part $\zeta_{ij}(t)$ represents the refractory effect of the ijth neuron at time t. They have different effects to determine firing of a chaotic neuron in the algorithm.

The first part $\xi_{ij}(t)$ is defined as follows:

$$\xi_{ij}(t+1) = \beta(E_{ij}(t) - \hat{E}), \tag{32}$$

$$E_{ij}(t) = \frac{1}{L}\sum_{k=1}^{L}\sum_{\omega\in\Omega} f_k(\omega)\log_2\frac{f_k(\omega)}{p(\omega)}, \tag{33}$$

where β (> 0) is the scaling parameter of the gain effect; $E_{ij}(t)$ is the objective function when a motif candidate position is moved to the jth position in the sequence s_i; $\hat{E}$, the entropy score of the current state. If a new motif candidate position is better than the current position, the quantity on the right-hand side of (32) becomes positive; a positive value leads to firing of the neuron. Thus, the gain effect encourages firing of the neuron, and such behavior is characteristic of a greedy algorithm.

However, the greedy algorithm gets stuck at a local minimum. To escape from the local minima, the refractory effect is assigned to each chaotic neuron. The second part $\zeta_{ij}(t)$ realizes the refractory effect. The refractory effect is one of the important properties of real biological neurons: once a neuron fires, a certain period of time must pass before the neuron can fire again. In the model of a chaotic neuron, the second part is expressed as follows:

$$\zeta_{ij}(t+1) = -\alpha\sum_{d=0}^{t} k_r^d x_{ij}(t-d) + \theta, \tag{34}$$

$$= -\alpha x_{ij}(t) + k_r\zeta_{ij}(t) + \theta(1 - k_r), \tag{35}$$

where α is the positive parameter; k_r, the decay parameter that takes values between 0 and 1; and θ, the threshold value. Thus, in (34), $\zeta_{ij}(t+1)$ expresses the refractory effect with the factor k_r because the more the neuron has fired in the past, the more negative is the first term on the right-hand side of (34).

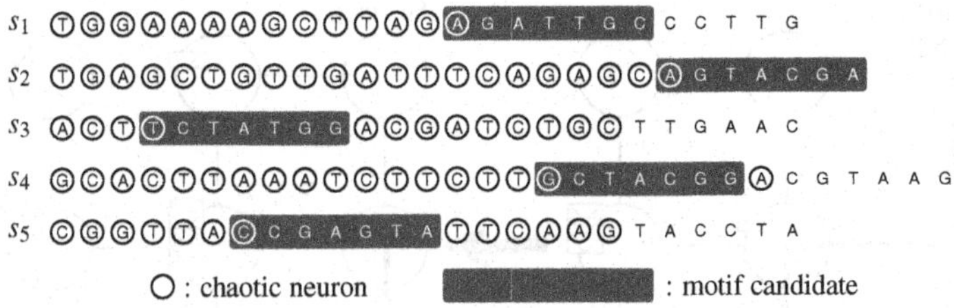

Fig. 6 Assignment of the chaotic neurons

This, in turn, reduces the value of $\zeta_{ij}(t+1)$ and causes the neuron to become a relatively resting state.

Figure 7 shows the results for real biological data set [72, 46, 47, 48, 49]. In Fig. 7, we show the average probabilities (%) of finding motifs in 50 trials. If the motifs are correctly found in 40 trials, the probability is $40/50 = 80\%$. In one trial, we change each motif candidate 500 times.

From Fig. 7(a), if β takes a small value ($\beta = 20$), the CMS shows low performance. Then, as the value of β increase ($\beta = 40$), the performances of the CMS becomes better (Fig. 7(b)). However, we cannot find motifs for too large values of β (Fig. 7(c)). The reason is that the CMS cannot escape from local minima if the strength of the greedy effect is stronger than that of the refractory effect. In other word, the searching approach is similar to steepest descent method.

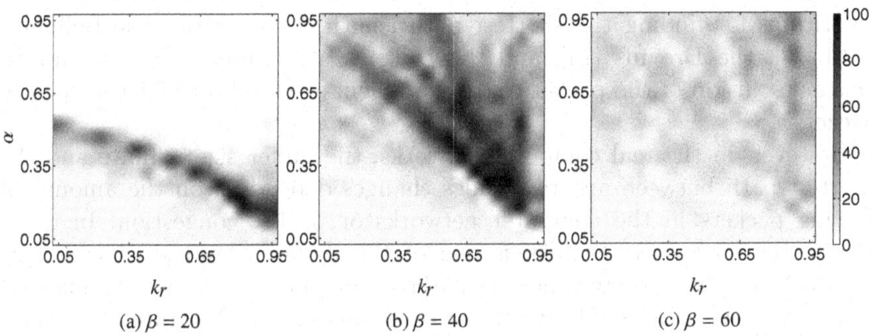

Fig. 7 Results of the chaotic motif sampler

3.3 Packet Routing Problems

A packet routing problem is one of the dynamical combinatorial optimization problems because the searching space dynamically changes depending on the state of the computer networks. In this section, as one of the applications of the chaotic neurodynamics to the dynamical combinatorial optimization problems, we explain a packet routing algorithm with chaotic neurodynamics for solving the packet routing problems [50, 51, 52, 53, 54, 55, 56, 57, 58].

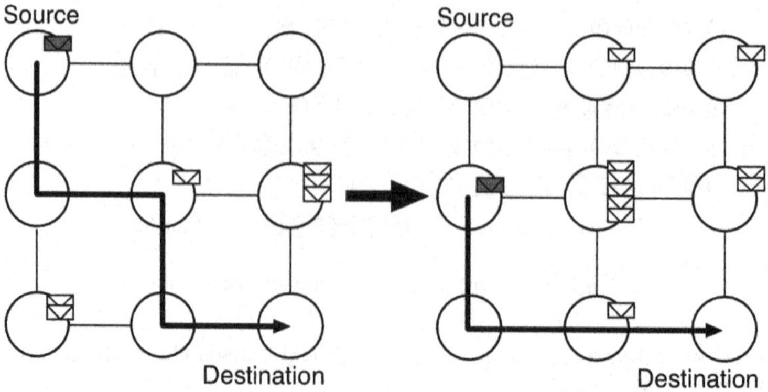

Fig. 8 An example of ideal computer networks. In this example, a gray packet is transmitted from a source to a destination. An arrow from the source to the destination expresses the shortest path of the gray packet. Because there are no packet congestion in the ideal computer network, we can easily find the shortest path of the gray packet. However, if we simply apply the basic strategy, such as the Dijkstra algorithm, to the real computer network, the packet congestion easily occurs

The packet routing problem is how to transmit the packets to their destinations as quickly and safely as possible depending on states of the computer networks. If a computer network is ideal, the buffer size of each node is infinite and throughputs of the nodes do not change. In such an ideal case, basic algorithms for finding the shortest path length, for example, the Bellman-Ford [73], the Dreyfus [74], and the Dijkstra algorithms [75], can find an optimal solution of the packet routing problem or the shortest paths for the packets.

However, in the real computer networks, the buffer size is finite and the shortest path between any two nodes changes depending on the amount of flowing packets in the computer networks or packet congestion. In other words, the computer network is one of the dynamic and stochastic networks [76, 77]. Because the shortest paths between nodes in the dynamic and stochastic networks are always changing depending on the state of the network, we have to consider how to avoid such congestion and how to transmit the packets securely and effectively by more sophisticated strategies.

Now, we define an objective function of the packet routing problem as follows;

$$r_i^* = \min_{j \in R_i} r_{ij} \quad (i = 1, \ldots, N_g), \tag{36}$$

where N_g denotes the number of existing packets; r_{ij} denotes the jth path from a source to a destination of the ith packet and it depends on a network state; R_i is the set of all possible paths r_{ij}. Equation (36) means to find r_i^* which is the shortest r_{ij} depending on the state of the computer networks.

The computer network model has N nodes, and the ith node has N_i adjacent nodes ($i = 1, \ldots, N$). Although there are several ways of how to assign a neural network to each node, we take the same way as [78, 79, 80, 50, 55].

To realize the packet routing algorithms with chaotic neurodynamics, first, we construct a basic neural network which functions to minimize a distance of a transmitting packet from the ith node to its destination. To realize this routing strategy, an internal state of the ijth neuron, ξ_{ij}, in the basic neural network is defined as

$$\xi_{ij}(t+1) = \beta \left(1 - \frac{d_{ij} + d_{jg(p_i(t))}}{d_c}\right), \tag{37}$$

where d_{ij} is the static distance of a path from the ith node to the jth adjacent node; $p_i(t)$ is a transmitted packet from the ith node at the tth iteration; $g(p_i(t))$ is a destination of $p_i(t)$; $d_{jg(p_i(t))}$ is the dynamic distance from the jth adjacent node to the destination of $p_i(t)$, that is, $d_{jd(p_i(t))}$ changes depending on $p_i(t)$; d_c is the diameter of the computer network; β is the normalization parameter which takes a positive value.

If the packets are transmitted to the destinations along only with the shortest paths, almost all the packets might be transmitted to the nodes through which many shortest paths pass. This behavior might lead to delay or lost packets. To avoid such an undesirable situation, one of the possible strategies is to memorize a node to which packets have just been transmitted for a while, and not to transmit the packets to the node. Then, we use a refractory effect peculiar to a chaotic neuron model [24]. The refractory effect is defined by

$$\zeta_{ij}(t+1) = -\alpha \sum_{d=0}^{t} k_r^d x_{ij}(t-d) + \theta, \tag{38}$$

where α is the positive control parameter of the refractoriness; k_r is the decay parameter of the refractoriness and takes between 0 and 1; θ is the threshold; $x_{ij}(t)$ is the output of the ijth neuron at time t which will be defined in (40).

Although the basic mechanism for the memory effect is realized by (37) and (38), mutual connections among neurons are also introduced to control firing rates of neurons, because too frequent firing often leads to a fatal situation of the packet routing. The internal state of the mutual connection effect is described as follows:

$$\eta_{ij}(t+1) = W - W \sum_{j=1}^{N_i} x_{ij}(t), \tag{39}$$

where W is a positive parameter and N_i is the number of adjacent nodes at the ith node. Because $W > 0$, if the number of firing neurons increases, then the second term of the right hand side becomes large, which again depresses the firing of the neuron at time $t+1$ and makes $\eta_{ij}(t+1)$ small.

Then, the output of the ijth neuron is finally defined by the sum of the above-introduced three internal states, $\xi_{ij}(t+1)$, $\zeta_{ij}(t+1)$, and $\eta_{ij}(t+1)$ as follows:

$$x_{ij}(t+1) = f\{\xi_{ij}(t+1) + \zeta_{ij}(t+1) + \eta_{ij}(t+1)\}, \tag{40}$$

where $f(y) = 1/(1+e^{-y/\epsilon})$, and ϵ is a positive but small parameter. In (40), if $x_{ij}(t+1)$ takes the value larger than $1/2$, the ijth neuron fires.

We compared the proposed chaotic routing strategy with a packet routing strategy using a neural network which has only the descent downhill dynamics of (37) (the descent downhill routing strategy) and a packet routing strategy using a tabu search (the tabu search routing strategy [25, 26]).

We conducted computer simulations of the packet routing for the scale-free networks [81]. Because real communication networks are scale-free [82], we adopted the scale-free topology as the network topology.

To evaluate performance of the three routing strategies, we used an arrival rate of the packets, and the number of packets arriving at their destinations. In this simulation, 20 scale-free networks of 100 nodes are prepared, and the quantitative measures, an arrival rate of the packets and the number of packets arriving at their destinations, are averaged over these 20 scale-free networks. Although we show only the results of the 100-node networks, we obtained the similar tendency for other 50-node and 200-node networks.

Results for the scale-free networks are shown in Fig. 9. In Fig. 9, the proposed chaotic routing strategy keeps the higher arrival rate of the packets than those of the descent downhill and the tabu search strategies for every packet generating probability (Fig. 9(a)). In addition, the chaotic routing strategy transmits more packets to their destinations than the descent downhill and the tabu search routing strategies for every packet generating probability (Fig. 9(b)). These results indicate that the chaotic neurodynamics is effective for avoiding the packet congestion by using the past routing history, which is realized by the refractory effect (38). Then, the chaotic routing strategy effectively routes the packets to their destinations without loss of the packets.

To reveal the effectiveness of the chaotic neurodynamics for the packet routing problems, we analyzed the network behavior of the chaotic routing strategy and the descent downhill routing strategy by spatial firing rates of neurons. The spatial firing rates by the routing strategies are shown in Fig. 10, which demonstrates that neurons in the chaotic routing strategy (Fig. 10(a)) are firing more uniformly than in the descent downhill routing strategy (Fig. 10(b)). Figure 10 shows that many paths for the packets are selected by the chaotic neurodynamics because the neurons in the chaotic routing strategy are uniformly firing during the simulations. As a result, the chaotic routing strategy transmits many packets to their destinations by selecting the transmission paths for the packets effectively.

The effectiveness of the chaotic neurodynamics for avoiding congestion of packets is analyzed using the method of surrogate data [54, 57]. In addition,

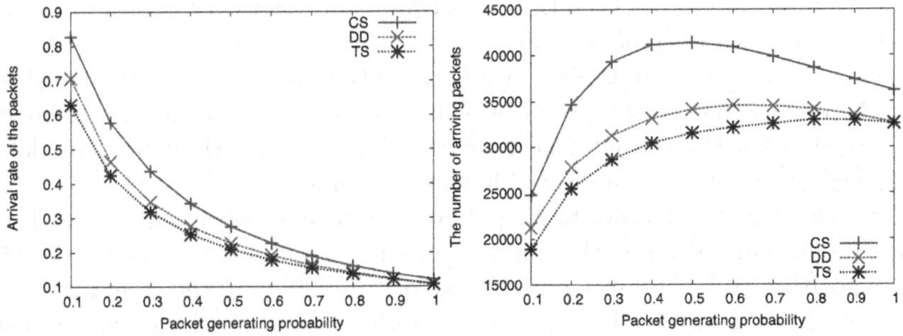

(a) (b)

Fig. 9 Relationship between the packet generating probability and (**a**) the arrival rate of the packets, and (**b**) the number of packets arriving at their destinations

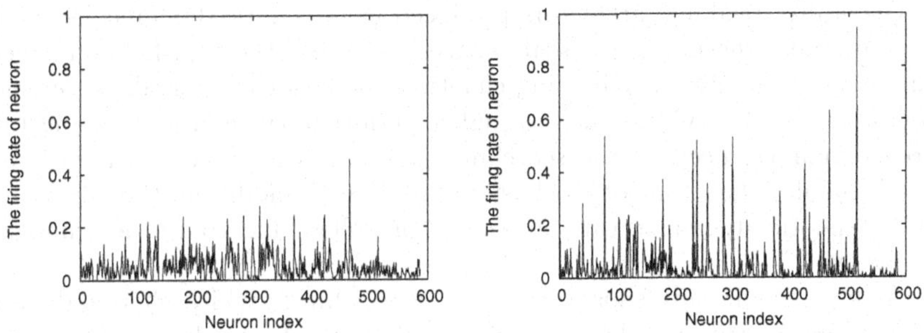

(a) (b)

Fig. 10 The spatial firing rates of neurons by (**a**) the chaotic routing strategy and (**b**) the descent downhill routing strategy. The packet generating probabilities in these figures are set to 0.4

the above-mentioned routing strategy with chaotic neurodynamics is modified by adding the waiting times until which the packets are transmitted at the adjacent nodes [56,58]. The results show that the modified chaotic routing strategy exhibits higher performance as compared to the conventional routing strategy with chaotic neurodynamics [51, 52, 53, 54, 57] and Echenique's algorithm [83, 84].

4 Conclusions

In this chapter, we have reviewed three methods for solving combinatorial optimization problems by using chaotic dynamics.

The first algorithm uses chaotic time series as additive dynamical noise that is injected to descent downhill dynamics of the recurrent neural network, or the Hopfield-Tank neural network. In this case, chaotic sequences are used to shake internal states of the Hopfield-Tank neural network in order to avoid undesirable traps into local minima.

The second algorithm is also based on the Hopfield-Tank neural networks, but the chaotic neuron proposed by Aihara et al. is used as a basic element. In this method, the refractoriness produced by the chaotic neuron model is effectively used to avoid undesirable local minima.

The third method used chaotic dynamics to control executions of local search algorithm, such as the 2-opt algorithm, the 3-opt algorithm, the Or-opt algorithm [19], the Lin-Kernighan algorithm [20], and the stem-and-cycle ejection chain method [21,22]. In this chaotic method [14,15,16,17,18], execution of the local search algorithm is encoded by firing of the chaotic neuron. Once a chaotic neuron fires, the firing of this chaotic neuron is inhibited for a while by the refractoriness, which restricts frequent firing of the chaotic neuron, or frequent execution of the same local search algorithm. Thus, the chaotic search can escape from local minima efficiently.

Generally speaking, attractors produced from chaotic dynamical systems have fractal structure in the state space, which has the zero Lebesgue measure. Thus, effective search using chaotic dynamics can be realized on such fractal attractors, which leads to higher performance than those using random dynamics, because the searching space of such fractal attractors are much smaller than that of stochastic search [8]. In addition, the algorithms using chaotic dynamics can be easily controlled due to its deterministic property.

As we have already shown, the third algorithm in which chaotic dynamics controls execution of local searches exhibits the best solving performance among the three methods. One of the key factors so that the third algorithm shows the highest performance is that chaotic search is realized with the refractory effect, or an exponential decay of the tabu effect. Moreover, the algorithm with chaotic dynamics can be easily implemented by analog circuits, which can drastically reduce the computational time to obtain good solutions. One of the limitations of the chaotic searching methods is that we have to tune parameters in the algorithms. However, this drawback can be resolved by developing an automatic parameter-tuning method based on analyses and controls of chaotic dynamics.

We have also reviewed applications of the third method to the vehicle routing problems [44,43,45] , the motif extraction problems [68,46,47,48,69, 49,70] , and the packet routing problems [51,52,53,54,57] . The results of computer simulations clearly show that the chaotic dynamics is very effective to solve these real-world application problems.

Although we have only discussed and showed the efficiency of the chaotic methods by computer simulations in this chapter, one of the most important research directions is to implement these algorithms by analog circuits. By the analog-circuit implementation of the chaotic search methods described in this chapter, we could develop a novel frontier of information processing that is based on a new computation principle by such nonlinear analog circuits [85, 86, 87, 88, 89, 90, 91, 92, 93, 94, 95, 96, 97, 98, 99, 100, 101, 102, 103].

The authors would like to thank Y. Horio, L. Chen, N. Ichinose, K. Umeno, T. Yamada, K. Sato, T. Hoshino and S. Motohashi for their valuable comments and discussions. This research is partially supported by the Japan Society for the Promotion of Science (JSPS) through its "Funding Program for World-Leading Innovative R&D on Science and Technology (FIRST Program)."

References

1. Hopfield, J.J., Tank, D.W.: 'Neural' computation of decisions in optimization problems. Biol. Cybern. 52, 141–152 (1985)
2. Hayakawa, Y., Marumoto, A., Sawada, Y.: Effects of the chaotic noise on the performance of a neural network model for optimization problems. Phys. Rev. E 51, 2693–2696 (1995)
3. Hasegawa, M., Ikeguchi, T., Matozaki, T., Aihara, K.: An analysis of additive effects of nonlinear dynamics for combinatorial optimization. IEICE Trans. Fundamentals E80-A(1), 206–213 (1997)
4. Hasegawa, M., Umeno, K.: Solvable performances of optimization neural networks with chaotic noise and stochastic noise with negative autocorrelation. In: Ishikawa, M., Doya, K., Miyamoto, H., Yamakawa, T. (eds.) ICONIP 2007, Part I. LNCS, vol. 4984, pp. 693–702. Springer, Heidelberg (2008)
5. Uwate, Y., Nishio, Y., Ueta, T., Kawabe, T., Ikeguchi, T.: Performance of chaos and burst noises injected to the hopfield nn for quadratic assignment problems. IEICE Trans. Fundamentals E87-A(4), 937–943 (2004)
6. Minami, Y., Hasegawa, M.: Analysis of relationship between solvable performance and cross-correlation on optimization by neural network with chaotic noise. Journal of Signal Processing 13(4), 299 (2009)
7. Nozawa, H.: A neural network model as a globally coupled map and applications based on chaos. Chaos 2(3), 884–891 (1992)
8. Chen, L., Aihara, K.: Chaotic simulated annealing by a neural network model with transient chaos. Neural Networks 8(6), 915–930 (1995)
9. Yamada, T., Aihara, K.: Nonlinear neurodynamics and combinatorial optimization in chaotic neural networks. Journal of Intelligent and Fuzzy Systems 5, 53–68 (1997)
10. Hasegawa, M., Ikeguchi, T., Motozaki, T., Aihara, K.: Solving combinatorial optimization problems by nonlinear neural dynamics. In: Proc. of IEEE International Conference on Neural Networks, pp. 3140–3145 (1995)
11. Wang, L., Li, S., Tian, F., Fu, X.: A noisy chaotic neural network for solving combinatorial optimization problems: Stochastic chaotic simulated annealing. IEEE Trans. on Systems, MAN, and Cybernetics - Part B: Cybernetics 34(5), 2119–2125 (2004)
12. Wang, L.P., Li, S., Tian, F.Y., Fu, X.J.: Noisy chaotic neural networks with variable thresholds for the frequency assignment problem in satellite communications. IEEE Trans. on Systems, MAN, and Cybernetics - Part C: Applications and Reviews 38(2), 209–217 (2008)

13. Wang, L., Liu, W., Shi, H.: Delay-constrained multicast routing using the noisy chaotic neural network. IEEE Trans. on Computers 58(1), 82–89 (2009)
14. Hasegawa, M., Ikeguchi, T., Aihara, K.: Combination of chaotic neurodynamics with the 2-opt algorithm to solve traveling salesman problems. Phys. Rev. Lett. 79(12), 2344–2347 (1997)
15. Hasegawa, M., Ikeguchi, T., Aihara, K.: Solving large scale traveling salesman problems by chaotic neurodynamics. Neural Networks 15(2), 271–283 (2002)
16. Hasegawa, M., Ikeguchi, T., Aihara, K.: Tabu search for the traveling salesman problem and its extension to chaotic neurodynamical search. Proceedings of International Symposium of Artificial Life and Robotics 1, 120–123 (2002)
17. Motohashi, S., Matsuura, T., Ikeguchi, T., Aihara, K.: The lin-kernighan algorithm driven by chaotic neurodynamics for large scale traveling salesman problems. In: Alippi, C., Polycarpou, M., Panayiotou, C., Ellinas, G. (eds.) ICANN 2009. LNCS, vol. 5769, pp. 563–572. Springer, Heidelberg (2009)
18. Matsuura, T., Ikeguchi, T.: Chaotic search for traveling salesman problems by using 2-opt and or-opt algorithms. In: Kůrková, V., Neruda, R., Koutník, J. (eds.) ICANN 2008, Part II. LNCS, vol. 5164, pp. 587–596. Springer, Heidelberg (2008)
19. Or, I.: Traveling salesman-type combinatorial problems and their relation to the logistics of regional blood banking. Ph.D thesis. Department of Industrial Engineering and Management Science, Northwestern University, Evanston, Illinois (1967)
20. Lin, S., Kernighan, B.: An effective heuristic algorithm for the traveling-salesman problem. Operations Research 21, 498–516 (1973)
21. Glover, F.: New ejection chain and alternating path methods for traveling salesman problems. Computer Science and Operations Research, 449–509 (1992)
22. Rego, C.: Relaxed tours and path ejections for the traveling salesman problem. European Journal of Operational Research 106, 522–538 (1998)
23. Aihara, K.: Chaotic neural networks. In: Kawakami, H. (ed.) Bifurcation Phenomena in Nonlinear System and Theory of Dynamical Systems, pp. 143–161. World Scientific, Singapore (1990)
24. Aihara, K., Takabe, T., Toyoda, M.: Chaotic neural networks. Phys. Lett. A 144, 333–340 (1990)
25. Glover, F.: Tabu search–part I. ORSA Journal on Computing 1, 190–206 (1989)
26. Glover, F.: Tabu search–part II. ORSA Journal on Computing 2, 4–32 (1990)
27. Matsuura, T., Ikeguchi, T.: An effective chaotic search by using 2-opt and Or-opt algorithms for traveling salesman problem. In: Proceedings of International Symposium on Nonlinear Theory and its Applications, vol. 21, pp. 77–80 (2008)
28. Matsuura, T., Ikeguchi, T.: A new chaotic algorithm for solving traveling salesman problems by using 2-opt and Or-opt algorithms. Technical Report of IEICE 107(561), 37–42 (2008)
29. Motohashi, S., Matsuura, T., Ikeguchi, T.: Chaotic search method using the Lin-Kernighan algorithm for traveling salesman problems. In: Proceedings of International Symposium on Nonlinear Theory and its Applications, pp. 144–147 (2008)

30. Motohashi, S., Matsuura, T., Ikeguchi, T.: The extended chaotic search method using the Lin-Kernighan algorithm. In: Proceedings of IEICE Society Conference, vol. A-2-5, p. 32 (2008)
31. Motohashi, S., Matsuura, T., Ikeguchi, T.: Parameter tuning method of chaotic search method for solving traveling salesman problems. In: Proceedings of IEICE General Conference, vol. A-2-14 (2009)
32. Motohashi, S., Matsuura, T., Ikeguchi, T.: Chaotic search based on the ejection chain method for traveling salesman problems. In: Proceedings of International Symposium on Nonlinear Theory and its Applications, pp. 304–307 (2009)
33. Kantz, H., Schreiber, T.: Nonlinear time series analysis. Cambridge University Press, Cambridge (2003)
34. Mazzini, G., Rovatti, R., Setti, G.: Interference minimization by autocorrelation shaping in asynchronous DS-CDMA systems: chaos-based spreading is nearly optimal. Electronics Letters 14, 1054 (1999)
35. Umeno, K., Yamaguchi, A.: Construction of optimal chaotic spreading sequence using lebesgue spectrum filter. IEICE Trans. Fundamentals 4, 849–852 (2002)
36. Aihara, K.: Chaos engineering and its application to parallel distributed processing with chaotic neural networks. Proc. of the IEEE 90(5), 919–930 (2002)
37. Adachi, M., Aihara, K.: Associative dynamics in a chaotic neural network. Neural Networks 10(1), 83–98 (1997)
38. Hasegawa, M.: Chaotic neurodynamical approach for combinatorial optimization. Doctor's thesis, Tokyo University of Science (2000)
39. Hasegawa, M., Ikeguchi, T., Aihara, K.: A novel approach for solving large scale traveling salesman problems by chaotic neural networks. In: Proc. of International Symposium on Nonlinear Theory and its Applications, pp. 711–714 (1998)
40. Hasegawa, M., Ikeguchi, T., Aihara, K.: Harnessing of chaotic dynamics for solving combinatorial optimization problems. In: Proc. of International Conference on Neural Information Processing, pp. 749–752 (1998)
41. Hasegawa, M., Ikeguchi, T., Aihara, K.: Exponential and chaotic neurodynamical tabu searches for quadratic assignment problems. Control and Cybernetics 29(3), 773–788 (2000)
42. Hasegawa, M., Ikeguchi, T., Aihara, K., Itoh, K.: A novel chaotic search for quadratic assignment problems. European J. of Operational Research 139, 543–556 (2002)
43. Hoshino, T., Kimura, T., Ikeguchi, T.: Two simple local searches controlled by chaotic dynamics for vehicle routing problems with time windows. In: Proc. Metaheuristics Int. Conf., CD–ROM, Montreal, Canada, June 25-29 (2007)
44. Hoshino, T., Kimura, T., Ikeguchi, T.: A metaheuristic algorithm for solving vehicle rouging problems with soft time windows by chaotic neurodynamics. Transactions of IEICE J90-A(5), 431–441 (2007)
45. Hoshino, T., Kimura, T., Ikeguchi, T.: A new diversification method to solve vehicle routing problems using chaotic dynamics. In: Complexity, Applications of Nonlinear Dynamics, pp. 409–412. Springer, Heidelberg (2009)
46. Matsuura, T., Ikeguchi, T.: A chaotic search for extracting motifs from DNA sequences. In: Proceedings of 2005 RISP International Workshop on Nonlinear Circuits and Signal Processing, pp. 143–146 (March 2005)

47. Matsuura, T., Ikeguchi, T., Horio, Y.: Tabu search and chaotic search for extracting motifs from DNA sequences. In: Proceedings of the 6th Metaheuristics International Conference, pp. 677–682 (August 2005)
48. Matsuura, T., Ikeguchi, T.: Refractory effects of chaotic neurodynamics for finding motifs from DNA sequences. In: Corchado, E., Yin, H., Botti, V., Fyfe, C. (eds.) IDEAL 2006. LNCS, vol. 4224, pp. 1103–1110. Springer, Heidelberg (2006)
49. Matsuura, T., Ikeguchi, T.: Statistical analysis of output spike time-series from chaotic motif sampler. In: Ishikawa, M., Doya, K., Miyamoto, H., Yamakawa, T. (eds.) ICONIP 2007, Part I. LNCS, vol. 4984, pp. 673–682. Springer, Heidelberg (2008)
50. Kimura, T., Nakajima, H., Ikeguchi, T.: A packet routing method for a random network by a stochastic neural network. In: Proceedings of International Symposium on Nonlinear Theory and its Applications, pp. 122–125 (2005)
51. Kimura, T., Ikeguchi, T.: A packet routing method using chaotic neurodynamics for complex networks. In: Kollias, S.D., Stafylopatis, A., Duch, W., Oja, E. (eds.) ICANN 2006. LNCS, vol. 4132, pp. 1012–1021. Springer, Heidelberg (2006)
52. Kimura, T., Ikeguchi, T.: Chaotic dynamics for avoiding congestion in the computer network. In: Corchado, E., Yin, H., Botti, V., Fyfe, C. (eds.) IDEAL 2006. LNCS, vol. 4224, pp. 363–370. Springer, Heidelberg (2006)
53. Kimura, T., Ikeguchi, T.: Optimization for packet routing using a chaotic neurodynamics. In: Proceedings of IEEE International Symposium on Circuits and Systems (2006)
54. Kimura, T., Ikeguchi, T.: A new algorithm for packet routing problems using chaotic neurodynamics and its surrogate analysis. Neural Computation and Applications 16, 519–526 (2007)
55. Kimura, T., Nakajima, H., Ikeguchi, T.: A packet routing method for complex networks by a stochastic neural network. Physica A 376, 658–672 (2007)
56. Kimura, T., Ikeguchi, T.: An efficient routing strategy with load-balancing for complex networks. In: Proceedings of International Symposium on Nonlinear Theory and its Applications, pp. 31–34 (2007)
57. Kimura, T., Ikeguchi, T.: An optimum strategy for dynamic and stochastic packet routing problems by chaotic neurodynamics. Integrated Computer-Aided Engineering 14(4), 307–322 (2007)
58. Kimura, T., Ikeguchi, T.: Chaotic routing on complex networks. IEICE Technical Report, pp. 25–30 (2007)
59. Kirkpatrick Jr., S., Gelatt, C.D., Vecchi, M.P.: Optimization by simulated annealing. Science 220, 671–680 (1983)
60. Rego, C.: Relaxed tours and path ejections for the traveling salesman problem. European Journal of Operational Research 106, 522–538 (1998)
61. TSPLIB, `http://www.iwr.uni-heidelberg.de/groups/comopt/software/TSPLIB95/`
62. Taillard, E., Guertin, F., Badeau, P., Gendreau, M., Potvin, J.Y.: A tabu search heuristic for the vehicle routing problem with soft time windows. Transportation Science 31, 170–186 (1997)
63. Bräysy, O., Gendreau, M.: Tabu search heuristics for the vehicle routing problem with time windows. Internal Report STF42 A01022 (2001)

64. Hoshino, T., Kimura, T., Ikeguchi, T.: Solving vehicle routing problems with soft time windows using chaotic neurodynamics. Tech. Rep. IEICE 105(675), 17–22 (2007)
65. Solomon's benchmark problems, `http://web.cba.neu.edu/~msolomon/problems.htm`.
66. Gehring and Homberger benchmark problems, `http://www.sintef.no/static/am/opti/projects/top/vrp/benchmarks.html`
67. Bräysy, O.: A reactive variable neighborhood search for the vehicle routing problem with time windows. INFORMS Journal on Computing 15(4), 347–368 (2003)
68. Matsuura, T., Anzai, T., Ikeguchi, T., Horio, Y., Hasegawa, M., Ichinose, N.: A tabu search for extracting motifs from DNA sequences. In: Proceedings of 2004 RISP International Workshop on Nonlinear Circuits and Signal Processing, pp. 347–350 (March 2004)
69. Matsuura, T., Ikeguchi, T.: Analysis on memory effect of chaotic dynamics for combinatorial optimization problem. In: Proceedings of the 6th Metaheuristics International Conference (2007)
70. Matsuura, T., Ikeguchi, T.: Chaotic motif sampler: Discovering motifs from biological sequences by using chaotic neurodynamics. In: Proceedings of International Symposium on Nonlinear Theory and its Applications, pp. 368–371 (2009)
71. Hernandez, D., Gras, R., Appel, R.: Neighborhood functions and hill-claiming strategies dedicated to the generalized ungapped local multiple alignment. European Journal of Operational Research 185, 1276–1284 (2005)
72. Lawrence, C.E., Altschul, S.F., Boguski, M.S., Liu, J.S., Neuwqld, A.F., Wootton, J.C.: Detecting subtle sequence signals: A gibbs sampling strategy for multiple alignment. Science 262, 208–214 (1993)
73. Bellman, E.: On a routing problem. Quarterly of Applied Mathematics 16, 87–90 (1958)
74. Dreyfus, S.: An appraisal of some shortest path algorithms. Operations Research 17, 395–412 (1969)
75. Dijkstra, E.W.: A note on two problems in connection with graphs. Numerical Mathematics 1, 269–271 (1959)
76. Fu, L., Rilett, L.R.: Expected shortest paths in dynamic and stochastic traffic networks. Transportation Research, 499–516 (1998)
77. Davies, C., Lingras, P.: Genetic algorithms for rerouting shortest paths in dynamic and stochastic networks. European Journal of Operational Research 144, 27–38 (2003)
78. Horiguchi, T., Ishioka, S.: Routing control of packet flow using a neural network. Physica A 297, 521–531 (2001)
79. Kimura, T., Nakajima, H.: A packet routing method using a neural network with stochastic effects. IEICE Technical Report 104(112), 5–10 (2004)
80. Horiguchi, T., Hayashi, K., Tretiakov, A.: Reinforcement learning for congestion-avoidance in packet flow. Physica A 349, 329–348 (2005)
81. Barabási, A.-L., Albert, R.: Emergence of scaling in random networks. Science 286, 509–512 (1999)

82. Faloutsos, M., Faloutsos, F., Faloutsos, C.: On power-law relationships of the internet topology. In: Proc. of the Conference on Applications, Technologies, Architectures, and Protocols for Computer Communication, pp. 251–262 (1999)
83. Echenique, P., Gomez-Gardens, J., Moreno, Y.: Improved routing strategies for internet traffic delivery. Physical Review E 70(056105), 325–331 (2004)
84. Echenique, P., Gomez-Gardens, J., Moreno, Y.: Dynamics of jamming transitions in complex networks. Europhysics Letters 71(2), 325–331 (2005)
85. Horio, Y., Aihara, K.: A large-scale chaotic neuro-computer. In: Proc. Joint Symp. for Advanced Science and Technology, Hatoyama, Japan, December 20, pp. 583–586 (1990)
86. Horio, Y., Suyama, K.: Switched-capacitor chaotic neuron for chaotic neural networks. In: Proc. IEEE Int. Symp. on Circuits and Syst., Chicago, IL, May 3-6, vol. 2, pp. 1018–1021 (1993)
87. Horio, Y., Suyama, K.: IC implementation of switched-capacitor chaotic neuron for chaotic neural networks. In: Proc. IEEE Int. Symp. on Circuits and Syst., London, UK, May 30-June 2, vol. 6, pp. 97–100 (1994)
88. Horio, Y., Suyama, K., Dec, A., Aihara, K.: Switched-capacitor chaotic neural networks for traveling salesman problem. In: Proc. INNS World Congress on Neural Networks, San Diego, CA, June 5-9, vol. 4, pp. 690–696 (1994)
89. Horio, Y., Suyama, K.: IC implementation of chaotic neuron and its application to synchronization of chaos. In: Proc. Int. Symp. on Nonlinear Theory and Its Applications, Ibusuki, October 6-8, pp. 185–188 (1994)
90. Horio, Y., Suyama, K.: Experimental observations of 2- and 3-neuron chaotic neural networks using switched-capacitor chaotic neuron IC chip. IEICE Trans. Fundamentals E78-A(4), 529–535 (1995)
91. Horio, Y., Suyama, K.: Dynamical associative memory using integrated switched-capacitor chaotic neurons. In: Proc. IEEE Int. Symp. on Circuits and Syst., Seattle, WA, April 30-May 2, pp. 429–435 (1995)
92. Horio, Y., Suyama, K.: Experimental verification of signal transmission using synchronized sc chaotic neural networks. IEEE Trans. Circuits Syst. I 42(7), 393–395 (1995)
93. Horio, Y., Kobayashi, I., Kawakami, M., Hayashi, H., Aihara, K.: Switched-capacitor multi-internal-state chaotic neuron circuit with unipolar and bipolar output functions. In: Proc. Int. Conf. on Microelectronics for Neural, Fuzzy, and Bio-Inspired Systems, Granada, Spain, April 7-9, pp. 267–274 (1999)
94. Horio, Y., Kobayashi, I., Kawakami, M., Hayashi, H., Aihara, K.: Switched-capacitor multi-internal-state chaotic neuron circuit with unipolar and bipolar output functions. In: Proc. IEEE Int. Symp. on Circuits and Syst., Orlando, FL, May 30-June 2, pp. 438–441 (1999)
95. Horio, Y., Aihara, K.: Chaotic neuro-computer. In: Chen, G., Ueta, T. (eds.) Chaos in Circuits and Systems, pp. 237–255. World Scientific, Singapore (2002)
96. Horio, Y., Aihara, K., Yamamoto, O.: Neuron-synapse IC chip-set for large-scale chaotic neural networks. IEEE Trans. Neural Networks 14(5), 1393–1404 (2003)
97. Horio, Y., Okuno, T., Mori, K.: Mixed analog/digital chaotic neuro-computer prototype: 400-neuron dynamical associative memory. In: Proc. of Int. Joint Conf. on Neural Networks, Budapest, Hungary, July 25-29, vol. 3, pp. 1717–1722 (2004)

98. Horio, Y., Okuno, T., Mori, K.: Switched-capacitor large-scale chaotic neurocomputer prototype and chaotic search dynamics. In: Proc. Int. Conf. on Knowledge-Based Intelligent Information and Engineering Systems, Wellington, New Zealand, September 20-24, vol. 1, pp. 988–994 (2004)
99. Horio, Y.: Analog computation with physical chaotic dynamics. In: Proc. RISP Int. Workshop on Nonlinear Circuit and Signal Processing, Hawaii, U.S.A., March 4-6, pp. 255–258 (2005)
100. Horio, Y., Ikeguchi, T., Aihara, K.: A mixed analog/digital chaotic neurocomputer system for quadratic assignment problems. INNS Neural Networks 18(5-6), 505–513 (2005)
101. Horio, Y., Aihara, K.: Physical chaotic neuro-dynamics and optimization. In: Proc. Int. Symp. on Nonlinear Theory and Its Applications, Bologna, Italy, September 11-14, pp. 449–502 (2006)
102. Horio, Y., Aihara, K.: Real-number computation through high-dimensional analog physical chaotic neuro-dynamics. In: Proc. Unconventional Computation: Quo Vadis?, Santa Fe, U.S.A., March 21-23 (2007)
103. Horio, Y., Aihara, K.: Analog computation through high-dimensional physical chaotic neuro-dynamics. Physica-D 237(9), 1215–1225 (2008)

Author Index